ESTIMATING AND TENDERING FOR CONSTRUCTION WORK

Martin Brook
BEng(Tech) FCIOB

Butterworth-Heinemann Ltd
Linacre House, Jordan Hill, Oxford OX2 8DP

᏶ A member of the Reed Elsevier group

OXFORD   LONDON   BOSTON
MUNICH   NEW DELHI   SINGAPORE   SYDNEY
TOKYO   TORONTO   WELLINGTON

First published 1993

British Library Cataloguing in Publication Data
Brook, Martin
  Estimating and Tendering for
  Construction Work
  I. Title
  692

  ISBN 0 7506 1531 1

Typeset by Vision Typesetting, Manchester
Printed and bound in Great Britain by Bath Press

Contents

Preface

My aims in this book are to introduce a practical approach to estimating and tendering from a contractor's point of view, and explain the estimator's role within the construction team. The book therefore differs from previous textbooks in three main ways:

1. In general it is assumed that it is the contractor who prepares estimates because in the majority of cases an estimate is produced to form the basis of a tender.
2. I have introduced many typical forms used by estimators to collate data and report to management. Most of the forms relate to two fictitious projects: a new lifeboat station and the construction of offices for Fast Transport Limited.
3. The pricing examples given in Chapter 11 have been produced using a typical build-up sheet. The items of work to which the prices relate are given at the top of each page. Estimating data are given for each trade so that students will have a source of information for building up rates. I suggest that before pricing exercises are undertaken, the first part of Chapter 11 should be read and an understanding of estimating methods should be gained from Chapter 5. The first pricing example is for a 'model rate' which gives a checklist of items to be included in a unit rate.

The estimating function has changed more in the last 10 years than at any time before. Many estimating duties can now be carried out by assistants using word-processors and computer-aided estimating systems. The estimator manages the process and produces clear reports for review by management.

Estimators need to understand the consequences of entering a contract, which is often defined by a complex combination of conditions and supporting documents. They also need to appreciate the technical requirements of a project from tolerances in floor levels to the design of concrete mixes, and from temporary electrical installations to piling techniques.

The Chartered Institute of Building publishes a series of guides to good practice – the Code of Estimating Practice and its supplements. I have not duplicated their fine work in this book but hope that my explanation and examples show how the guidelines can be used in practice.

Contractors now assume an active role in providing financial advice to their clients. The estimator produces financial budgets for this purpose and assembles cost allowances for use during construction. Computers have been introduced by most organizations, with a combination of general-purpose and specialist software. The benefits are difficult to measure during the tender period; but for the handover of successful tenders, adjustments can be made quickly, information can be presented clearly, and data can be transferred in a more compact form.

The changes brought about by the introduction of SMM7 and the other principles of Coordinated Project Information have reduced the number of items to be measured in a typical building contract. The item descriptions no longer provide information for pricing and the estimator must always refer to the specification and drawings. In practice this is time-consuming for both contractors and sub-contractors, and the amount of paperwork has increased immensely. Nevertheless, contractors welcome the widespread use of bills of quantities as a fair basis for preparing tenders and look forward to the time when there is a common approach to electronic data transfer.

I recognize and support the role of women in construction and ask readers to accept that the use of the masculine pronoun is intended to refer equally to both sexes.

Martin Brook
1993

Acknowledgements

I wish to acknowledge the help given by Michael Hawkridge of Christiani and Nielsen for checking the text and figures, and Jane Brook for the cartoons.

Figures

Abbreviations used in the text

BEC	Building Employers Confederation
BPF	British Property Federation
BS	British Standard
BWIC	Builders Work In Connection
CAD	Computer-Aided Design (or Draughting)
CAWS	Common Arrangement of Work Sections
CD	Compact Disc
CESMM3	Civil Engineering Standard Method of Measurement Third Edition
CI/SfB	Construction Index – Samarbetskommitten för Byggnadsfragor
CIOB	Chartered Institute of Building
COEP	Code of Estimating Practice (Published by the CIOB)
Conc	Concrete
CPSSST	Code of Procedure for Single Stage Selective Tendering
CPI	Coordinated Project Information
DOS	Disk operating System
DOT	Department of Transport
Exc	Excavation
FCEC	Federation of Civil Engineering Contractors
ICE	Institution of Civil Engineers
Inc	Included
JCT	Joint Contracts Tribunal
LCD	Liquid-Crystal Display
LOSC	Labour Only Sub-contractor
MB	Megabyte
ne	Not exceeding
NJCC	National Joint Consultative Committee for Building
PC	Prime Cost
PC	Personal Computer
PQS	Private Quantity Surveyor
Prov	Provisional
Quant	Quantity
RAM	Random-Access Memory
RIBA	Royal Institute of British Architects
RICS	Royal Institution of Chartered Surveyors
ROM	Read-Only Memory

SMM	Standard Method of Measurement
SMM6	Standard Method of Measurement of Building Works: Sixth Edition 1978
SMM7	Standard Method of Measurement of Building Works: Seventh Edition 1988

Organization of the estimating function

'The corporate image consultant is in reception, Sir.'

The role of the contractor's estimator is vital to the success of the organization. The estimator is responsible for predicting the most economic costs for construction in a way which is both clear and consistent. Although an estimator will have a feel for the prices in the marketplace, it is the responsibility of management to add an amount for general overheads, assess the risks and turn the estimate into a tender. The management structure for the estimating function tends to follow a common form with variations for the size of the company. In a small firm, the estimator might be expected to carry out some quantity surveying duties and will be involved in procuring materials and services. The estimating section in a medium-sized or large construction organization will often comprise a chief estimator, senior estimators and estimators at various stages of training. The larger estimating departments may have administrative and estimating assistants who can check calculations, photocopy extracts from the tender documents, prepare letters and enter data in a computer-assisted estimating system.

The estimating team for a proposed project has the estimator as its coordinator and is usually made up of a contracts manager, buyer, planning engineer and quantity surveyor. The involvement of other people will vary from company to company. A quantity surveyor is often consulted to examine amendments to conditions of contract, prepare a bill of quantities and identify possible difficulties which have been experienced on previous contracts. Clients sometimes like to negotiate agreements with quantity surveyors where follow-on work is to be based on pricing levels agreed for previous work. A planning engineer may be asked to prepare a preliminary programme so that the proposed contract duration can be checked for possible savings. He can also prepare method statements, temporary works designs, organizational charts and site-layout drawings. The buying office will provide valuable information leading to the most economic sources for the supply of materials and plant. In many organizations today, the buyer is responsible for getting quotations from suppliers and sub-contractors. At the very least, the buyer helps prepare lists of suitable suppliers, keeps a library of product literature and advises on likely price changes. Site managers should report on the technical and financial progress of their projects so that the estimator can learn from the company's experience on site.

The aim of the team is to gain an understanding of the technical, financial and contractual requirements of the scheme in order to produce a prediction of the cost of construction. The construction manager or director will then use the net cost estimate to produce the lowest commercial bid at which the company is prepared to tender. Figure 1.1 shows the various stages in preparing a tender and the action needed with successful tenders.

The cost of tendering for work in the construction industry is high and is paid for by part of the general overhead which has been added to each successful tender. The chief estimator needs to be sure there is a reasonable chance of winning the contract if the organization is in competition with others. The decision to proceed with a tender is based on many factors including the estimating resources available, extent of competition, tender period, quality of tender documents, type of work, location, current construction workload and conditions of contract. With all these points to consider, a chief estimator could be forgiven for declining many invitations to tender to maintain a high success rate and avoid uncompetitive bids which can lead to exclusion from approved lists. On the other hand, he must recognize the goodwill which often flows from submitting competitive prices and the need to carry out work which might lead to suitable and profitable contracts.

There are several forms which can be used to plan, control and monitor estimating workload. The first is a chart to show the opportunities to tender when they have been confirmed. The information for this programme usually comes

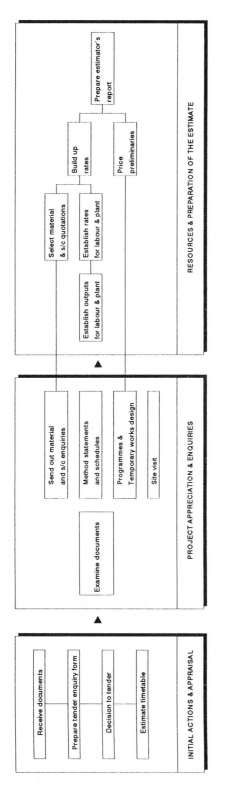

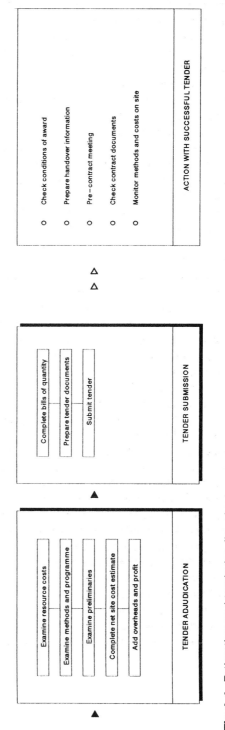

Fig 1.1 *Estimating and tendering flowchart*

from the marketing personnel who are responsible for bringing in invitations to tender for projects which are in line with company strategy. The chief estimator will prepare a bar chart (Fig. 1.2) to show how the estimators will be assigned to present and future tenders, showing the expected dates for receipt of documents and submission of tenders. Copies are sent to heads of other departments so they can plan their input; they may also wish to attend the adjudication meetings. A tender register is also needed (Fig. 1.3) to record the main details of each tender such as reference number, client, price, tender date and an analysis of performance in relation to the competition.

The success ratio for a construction firm is often quoted at about 1 in 6 although it can be as high as 1 in 10 and down to 1 in 4 where competition is limited. Since the directors of a company are more concerned with turnover, then success is better measured in terms of value, and the estimating department may be given annual targets to meet. Clearly, negotiated work can save a great deal of abortive tendering. The work flow in an estimating department is never constant; the ideal situation is to have people available who are multi-disciplinary and can deal with all administrative tasks. A buyer can provide an invaluable service in sending out enquiries and chasing quotations. His knowledge of local suppliers and current discounts is essential at the adjudication meeting when decisions need to be taken about the availability and future costs of materials.

Estimators are drawn from two sources; direct from school with some good grades in GCSE subjects which suggest a potential to study to a higher technician or professional level, or from experienced staff where management has identified an aptitude and willingness for the job. In both cases a reasonable time must be spent on site to gain experience in construction methods, materials identification, use and practice. The skills which are needed are the ability to read and interpret technical documentation, the ability to communicate with clients, specialists and other members of the team, and the faculty to make accurate calculations.

Technically an estimator must have a working knowledge of all the major trades, to foresee the time and resources which will be needed and to identify packages of work to be carried out by sub-contractors and the direct workforce. It is also necessary to have the ability to take off quantities from drawings, whether there is a bill of quantities or not, and to understand the elements already measured where bills of quantities are provided.

An estimator needs to refer to many information sources either in book form or through more modern means such as microfiche, on-line databases and CD-ROM. The following list shows some of the basic material required:

- Code of Estimating Practice (COEP)
- Code of Procedure for Single Stage Selective Tendering (CPSSST)
- Standard method of measurement (SMM)
- Standard forms of contract

| CB CONSTRUCTION LIMITED | ESTIMATING PROGRAMME | | | | | | | | | | | | | Date : 3.6.94 |

Estimator	Project title	June				July				August			
		6	13	20	27	4	11	18	25	1	8	15	22
JOHN EDWARDS	Colliery Office	XXXXXXXAXT											
	Lifeboat Station		XXXXXXXXXXXXXXXXXXXXAT			hhhhhhhhhhh							
JEAN SMITH	Access road	XXAT											
	Treatment works		XX XXXXXXXXXXXXAXT										
	Superstore				X XXXXXXXXAXXT		hhhhh						
GRAHAM THOMAS	Fast Transport	XXXXXXXXXXXXXXXXXXXXXXXXXXXXXXXAXT											

Key :   A = Adjudication   T = Tender date   h = holiday

Fig 1.2 *Estimating programme*

CB CONSTRUCTION LIMITED

TENDER REGISTER

Tender no	Tender title	Date rec'd	Location	Client	QS Architect Engineer	Tender details		Tender evaluation			
						Date	Price	No of tenderers	Rank	% over lowest	% over mean

Fig 1.3 *Tender register*

- Standard specifications for highways and water industries
- National Working Rule Agreement
- Definition of prime cost of daywork
- Daywork plant schedules
- Trade literature:
  - (a) standard price lists
  - (b) technical product information
- Trade directory of suppliers and sub-contractors
- Reference data for weights of materials

The work of the estimating department is an important part of the company's quality management system. The procedures adopted and decisions made at tender stage will often determine the way in which the project is carried out. To meet the objectives of a quality plan, the staff must document the way in which they intend to work and produce a manual or handbook which contains the standard procedures for preparing and submitting tenders. This document is particularly useful for newly appointed staff, not to change all their methods for pricing but to provide a framework for good practice in preparing estimates and consistent records and reports for others.

Many organizations developed procedures for estimating long before the more formalized approach brought about by BS 5750 registration. Quality assurance provides confidence that a product is fit for its purpose and is provided on time at the right price. It is not sensible to apply this approach to producing an estimate because it is always finished on time, the price is not always right and the terms have to comply with the instructions if the tender is to be considered. The estimating function is, however, part of the quality assurance system. The benefits are:

1. *Profitability* – an improvement to the profitability of the organization.
2. *Accuracy* – a reduction of errors.
3. *Competence* – better-trained staff.
4. *Efficiency* – work properly planned and systematically carried out.
5. *Job satisfaction* – for the whole estimating team.
6. *Client satisfaction* – leading to likelihood of repeat business.

2 *Procurement paths*

Introduction

The Banwell Report, published in 1964, expressed the view that existing contractual and professional conventions do not allow the flexibility which is essential to an industry in the process of modernization. The report of the committee asked the industry to experiment to secure efficiency and economy in construction.

The traditional method of organizing construction work starts with appointing a consultant designer, usually an architect or engineer, or both. Other specialists may be needed; in particular, a quantity surveyor is appointed to provide cost information, prepare bills of quantity, compare bids and maintain financial management during construction.

Since the early 1960s the construction industry has experienced significant changes in the way in which contracts are managed. In some cases contractors have been brought in at an early stage as full members of the design team, in others clients have appointed project managers to act on their behalf. During the 1980s clients have become increasingly concerned about problems such as poor design, inadequate supervision, delays and increased costs. They have also been critical of the separation of design from construction, particularly between the building professions.

In an attempt to overcome some of these long-standing criticisms, in 1983 the British Property Federation published its manual for building design and construction. They wanted to introduce a new system to change attitudes and alter the way in which the members of the construction team deal with one another. The BPF also tried to remove some of the overlap of effort between quantity surveyors and contractors without the need for the traditional bill of quantities.

The design and build method has gradually grown in popularity during the last three decades by offering single-point responsibilities, certainty of price and short overall durations. Management contracting has become popular for large complex projects although some people see construction management as a more attractive choice. An alternative, which is sometimes forgotten, is the client's own in-house design team, usually led by a project manager who supervises designers, cost specialists and contractors. In fact, this method accounts for a large part of construction work because it is the one commonly used in the public sector.

Clients' needs

Client organizations are divided between those in private and public sectors although this distinction is becoming more difficult to define since the privatization of many national bodies. The private sector includes industrial, commercial, social, charitable and professional organizations, and individuals. The public sector is taken to mean government departments, nationalized industries, statutory authorities, local authorities and development agencies. The experience which a client has of building procurement ranges from extensive, in the case of a client with a project management team, to none, where a private individual may want a development only once in a lifetime.

Clients will usually identify their needs in terms of commercial or social pressure to change; by an examination of primary objectives such as:

1. Space requirements: the need to improve production levels, add to production capacity, accommodate new processes or provide domestic or social accommodation;
2. Investment: to exploit opportunities to invest in buildings;
3. Identity: to enhance the individual's or organization's standing in its market or society;
4. Location: could lead to a better use of resources, capture a new market or improve amenity;
5. Politics: mainly in the public sector.

The client's experience of building will influence his expectation of the industry. Property developers, on the one hand, can influence their professional advisers and the contractual arrangements, and select a contractor with the right commitment to meeting project targets. The main aim is to achieve a degree of certainty in the building process. On the other hand, individuals and inexperienced clients are guided by their advisers and contractors, and will be offered what the construction team think they need.

In general a client aims to appoint a team which he can trust and rely on to reduce uncertainties during a building's design, construction and use. This is achieved by control of the following:

1. The design: by designing to a budget, taking advantage of the contractor's experience, avoiding excessive use of new systems, designing for buildability, safety, security, producing a good life expectancy and low maintenance, allowing flexibility for future change and employing environmental and energy-efficient designs;
2. The time: by contractors accepting more responsibility for meeting completion dates, and designers being more aware of the importance of complete information well in advance of work on site;
3. The cost: by achieving realistic cost estimates and tenders which reflect the final cost, reducing risk of contractual claims stemming from poor documentation and late receipt of information, and avoiding delays which can cause loss of revenue and costly funding arrangements.

Many clients are prepared to pay for a good service and see these objectives being met through alternative methods of contracting.

The client has traditionally occupied a passive role in the construction process. The standard forms of contract require the employer to pay for work properly executed, give possession to the site on the agreed date and appoint his professional team to design, supervise and inspect the work and account the finances. A more realistic view is that the client is the most important member of the team because as patron for the scheme he identifies the need for the building and he must pay everyone who is directly or indirectly involved in the construction process. This is why we now see clients taking a more active part in the control of construction work and in part explains the emergence of construction management in the UK.

Contractor involvement

During the late 1980s clients were looking for procurement methods which could quickly produce (or refurbish) large buildings with complex designs. Clearly the contractor needed to contribute to the design phase and continue to advise on the design during construction. At the same time, where projects were less complex, design and build systems were being adopted for both building and civil engineering projects.

In order to respond to these different needs, contractors have developed a wider range of construction services, sometimes setting up separate divisions within a company. The danger is that a contractor more used to working in a traditional market may fail to achieve the objectives expected by the clients. This can occur where there is a lack of trained staff and there is an over-reliance on specialists, for design and construction, who are sometimes engaged on onerous conditions of sub-contract. For construction management, as seen in the USA, to flourish, contractors must accept the responsibility for producing detailed drawings and cost-effective production techniques. Whichever method is used, there will usually be a number of tendering stages which encourage the parties to harmonize their aims and develop cooperation and trust, which did not always happen in the past. If this is the way ahead, then architects and quantity surveyors will concentrate on creating an outline of the client's requirements, providing financial advice and setting up independent monitoring systems on site.

In civil engineering, there are generally fewer professional interests, and an engineer, whether working for a client or contractor, works in a similar way. Civil engineers understand the standard documentation which is used for most engineering schemes. Contractors can, however, influence the design for civil engineering work significantly, and often submit tenders with alternative bids which can offer substantial savings to a client.

Apportionment of risk

The procurement system, and associated contractual arrangement, will dictate the financial and other risks borne by the parties to the contract. Risk cannot be eliminated by choosing a particular form of contract, but will be shifted towards one party or the other. A guide to how the risks are divided for each contractual arrangement is given in Fig. 2.1.

Lump-sum contracts based on complete pre-tender design and full documentation spread a smaller risk of cost overrun evenly between the parties. Results may be further improved by using a selective list of tenderers, avoiding nominations, checking ground conditions, and reducing the guesswork needed by contractors at tender stage. A contract where the price is calculated from a schedule of rates has two major problems:

1. The contractor is unable to identify the full extent of the work at tender stage and is thus unable to plan and accurately assess his overheads, and
2. The client will not know the full price of the work until the contract is complete.

A cost-reimbursement contract allows the contractor to claim all the prime cost of carrying out the work on an 'open book' basis and amounts are paid for site overheads and the management fee. Although this arrangement has the benefit of a quick start, there is little incentive to save time or costs. It would be unfair to say, however, that management contracting is more expensive than an alternative approach. All the works contracts are let competitively and the management fees

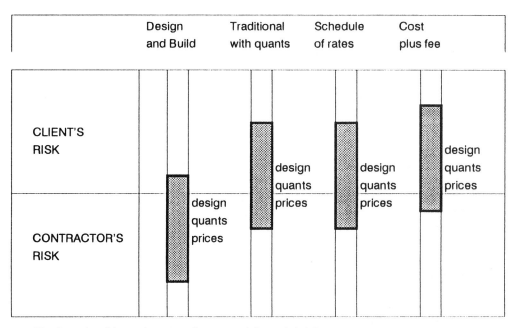

	Design and Build	Traditional with quants	Schedule of rates	Cost plus fee
CLIENT'S RISK		design quants prices	design quants prices	design quants prices
CONTRACTOR'S RISK	design quants prices	prices	prices	

Fig 2.1 *A guide to the apportionment of financial risk*

are surprisingly low. It has been suggested that the management contractor makes more money by looking after the payments to package contractors. This point is often made in acclaiming the benefits of construction management where the client deals directly with the specialist contractors.

Traditional method

The traditional structure for project procurement shown in Fig. 2.2 is seen as a sequential method because the employer takes his scheme to an advanced stage with his professional team before appointing a contractor. The consultant's role is seen as an independent one. The designer is employed to advise the client, design and ensure that the work is kept within the cost limit and that it complies with the standards required. A quantity surveyor can be engaged to give guidance on design costs and budgets, prepare bills of quantities, check tenders, prepare interim valuations and advise on the value of variations. Consultant structural and services engineers may be employed by either the client or his advisers to design the specialist parts of the project.

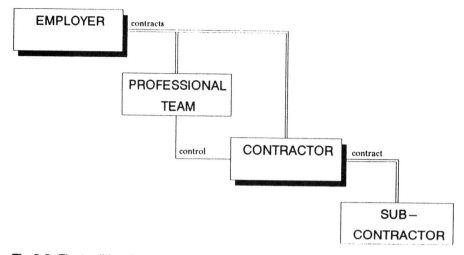

Fig 2.2 *The traditional procurement method*

Separating responsibilities for design and construction is seen as the main reason for the move to alternative contractual arrangements. The building industry suffers from the old distinctions between the professional interests and suspicion caused by ignorance of each others' work. In civil engineering there is more freedom for individuals to move to and from consultants and contractors' organizations – there is an understanding of each other's point of view.

Instead of the direct appointment of consultants, many major building owners and developers make use of in-house project managers either to control independent consultants or to carry out all the design and financial control of the project. Project management is therefore seen as a management tool and not a procurement system. The JCT Standard Form of Building Contract 1980 and JCT Intermediate Form of Contract 1984 with the Agreement for Minor Works are the most popular forms for building work. The ICE Form of Contract is used for most civil engineering work in both the public and private sectors, and GC/Works/1 is used for traditional civil engineering and building contracts let by central government departments. The continuing high sales of these contracts point to the commanding position of traditional methods.

Design and build

The design and build arrangement is an attractive option for clients. It simplifies the contractual links between the parties to the main contract (see Fig. 2.3) because the contractor accepts the responsibility for designing and constructing.

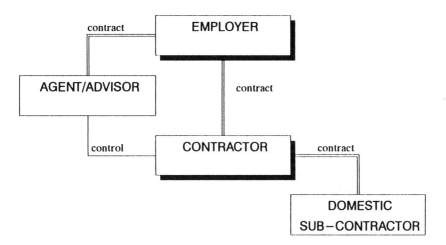

Fig 2.3 *The design and build system*

The benefits include single-point responsibility, prices which reflect more closely the final cost to the client, inherently more buildable designs and an overlap of design and construction phases leading to early completion. A distinction is sometimes drawn between design and build and a package deal, the latter being an agreement for the contractor to provide a semi-standardized or off-the-peg building which can be adapted to meet the client's needs. The contractual arrangement known as 'Turnkey' allows a client to commission from a single contractor all the requirements of a scheme in the shortest possible time. Apart from the usual design and construction responsibilities, the agreement will often include land acquisition, short- and long-term finance, commissioning, fitting out and recruitment and training of personnel.

The contractor may commission design and cost services from outside consultants or can employ a design team from within his own organization. Occasionally the client will ask the contractor to adopt a design started by his preferred consultant. Assigning (or 'novating') a design team in this way can arise when a client decides to switch to design and build from a traditional method. A designer needs a flexible outlook when he is thrust from the freedom of private practice, talking to a valued client, to being responsible mainly to the contractor, working to tight financial and time constraints, seeking solutions which satisfy the client's brief and enlarge the contractor's profits.

At tender stage the employer will introduce some competition, either open or selective tendering, which is followed by clarification of the agreement and negotiation. The National Joint Consultative Committee for Building published an advisory booklet in 1985 for private clients and public authorities planning to engage a contractor who would be responsible for the design and construction of a

15

building project. The Code of Procedure for Selective Tendering for Design and Build stresses the importance of full and clear documents setting out the Employer's Requirements. The number of contractors invited to submit tenders in the form of Contractor's Proposals should be limited to three or four firms to reduce the high tendering costs. Each firm invited to tender for design and build work is carefully selected not only for its financial standing and construction record but also for its design capability and management structure for the work.

The Code of Procedure recognizes the need for longer tendering periods (often three to four months and longer on more complex schemes) and where extensive specialist work or negotiations with statutory bodies is required even more time may be needed. The employer must clearly state the form and content of the contractor's proposals and say whether the price alone will determine the offer accepted. The Code suggests that the design proposals and contract sum analysis are supporting documents which could be submitted separately. The contractor's proposals must be checked with great care, because if there is a discrepancy in the employer's requirements the contractor's proposals will prevail, without any adjustment to the contract sum.

Before entering a design and build arrangement a client should consider the drawbacks. A contractor may offer a functional design which is not aesthetically appealing; he is inclined to develop a low-cost design with opportunities to increase his margins. A contractor might make a client's brief fit his own preferred solution; the long-term life of a building might be overlooked and if the brief is vague, the client could pay an inflated price or take possession of an inferior building. A client may not realize the importance of independent professional advice. The cost of abortive designs and tendering is a heavy burden on contractors' overheads and eventually the costs will be passed on to clients.

It would be difficult to support these criticisms now that design and build is so well established. Professional contractors have taken a pride in their approach to this system, which reduces conflict between the parties and gives the client single-point responsibility for design, time and cost.

In 1981 the Joint Contracts Tribunal published a new form of building contract with contractor's design, and an addendum for changing existing standard forms where the contractor must prepare the design for only part of the works. The new form was based on the 1980 standard form of building contract, with quantities. The contract is for a lump-sum price payable in stages or monthly. In place of a bill of quantities the form provides for a contract sum analysis to assist those preparing interim valuations and valuing variations. It must be said, however, that the contract sum analysis only helps with significant variations and is of no use with day-to-day changes. The JCT published Practice Note CD/1B in 1984 which includes a useful explanation about the purpose and recommended structure for the contract sum analysis.

Management contracting

During the 1980s clients were attracted to management contracting because it offered early starts to large-scale and often complex construction projects. The management contractor is appointed to work with the professional team, to contribute his construction expertise to the design and later to manage the specialist 'package' or 'works' contractors. He is responsible for the smooth running of the work on site so that the contract can be finished within time and cost. Although most of the major contractors have undertaken work using management contracts, there has been a feeling that it is not a final solution and a better method will evolve in the future. One development has been a combination of design and build and management contracting whereby the contractor produces a design and guaranteed maximum price and the work is later assigned to a number of major package contractors.

Figure 2.4 shows the system adopted by the Joint Contracts Tribunal in 1987 for its Standard Form of Management Contract. The contract was needed to meet the growth in the market and the need for standard documentation to replace the many improvised forms which had been used. A management contractor is selected using the following criteria:

1. Experience of management contracting.
2. Quality and experience of project staff.
3. Fee.
4. Programme and method statement.

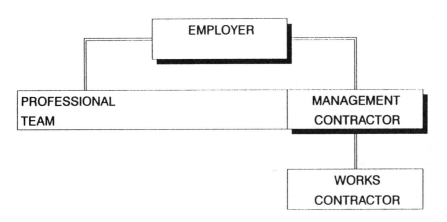

Fig 2.4 *Management contracting*

The consultants, grouped under the title of 'the professional team', prepare the drawings, specifications and bills of quantities for the various works contracts. The Architect (or Contract Administrator) leads the Professional Team and issues instructions to the management contractor on behalf of the employer. The management contractor's role is in coordinating the design and preparing cost studies at the pre-construction stage. During construction his duties include placing and letting contracts with specialists, cost studies, setting out, provision of shared facilities, plant, and scaffolding, planning and monitoring the work, and coordinating all the activities on site, but not carrying out the permanent works. The management contractor's main duty is to cooperate with the professional team in the above functions.

The JCT Management Contract is not a lump-sum contract. The Employer pays the prime cost of carrying out the work and the fees for providing the management services. These fees will be either a lump sum or calculated as a percentage of the contract value. The recommended retention is 3% applied to both the management and works contracts, but not the fee because fees are calculated after retention is deducted. Trade discounts including the 2.5% contractor's discount are deducted from the management contractor to the benefit of the employer.

Clients are attracted to management contracting for the following reasons:

1. Construction can start before design is complete, and design can be changed during the construction phase;
2. Construction expertise is available to improve on the design;
3. Better coordination of specialist contractors through detailed planning of work packages and common facilities;
4. A contractor's knowledge of construction costs is used to maintain tight budgetary control.

Some problems have emerged, and management contracting shows signs of decline, except in very large projects. Contractors are less enthusiastic now that margins have fallen and sub-contractors have demanded prompt payments. For works contractors the conditions of contract are becoming more demanding with regard to the management contractor's right of set-off, liquidated damages, performance bonds and guarantees. The specialists often carry the burden of late changes to drawings and specifications which are more common when design development takes place during construction. The client cannot be sure of the final cost and will carry more risk. This is because the management contractor can pass on all the costs incurred for each trade, site staff and site facilities.

Construction management

In the USA, where the roles of the professionals are different, the client or his project manager will take a more active part in the construction phase. A construction manager is appointed as a professional consultant with powers to inspect work on site and issue instructions (see Fig. 2.5).

The client has a greater control over funds during construction because he has a contract with all the trade and specialist contractors. These contractors welcome the direct links with the client partly for the higher status this brings but, more importantly, because the lines of communications are clearer and payments are made sooner.

In the UK, some clients would not want to deal directly with sub-contractors or be involved in every problem of time and cost that could arise. A standard contract is not yet available for this system and a manager agreement has not been drafted. Nevertheless, there is evidence of experimentation by large property developers who want firm control of and involvement in building projects.

There have been some spectacular building failures in the USA; a congressional inquiry in 1984 found that design quality can be impaired by excessive speed and cost-cutting exercises. Problems have been found when designers, who are often selected on a least-fee basis, pass preliminary designs to works contractors who produce the detailed drawings. This is a division of responsibilities which can lead to errors and legal action.

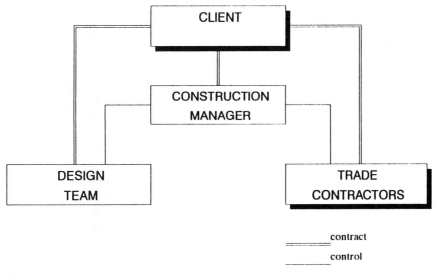

Fig 2.5 *Construction management*

Forms of contract

Introduction

Standard forms of contract exist to identify the roles and responsibilities of the parties, and their agents; and provide rules to protect and direct the parties should things go wrong. Clients have a wide choice of standard contracts for construction work, in particular the forms used for building which cover most of the common procurement systems. Standard conditions have been written by bodies such as the Joint Contracts Tribunal and Institution of Civil Engineers, following changing procurement methods in the industry – they seldom lead. The alternative approach would be to produce a common form of contract for all construction work whether in the public or private sector, building or civil engineering, English or Scottish law. This idea is not new – it was one of the principal recommendations of the Banwell Report in 1964.

Where a standard form of contract is proposed, an estimator must carefully examine the information which will be inserted in the Appendix and note any amendments to the standard conditions so that the terms of the offer can be evaluated. An estimator should assess the cost of complying with certain terms and advise management of any onerous conditions which may influence the bid. Non-standard forms of contract are sent to the commercial department, company secretary or director so that the conditions can be evaluated before the adjudication meeting.

Essentials of a valid construction contract

Construction contracts are the same as any other contract, and in the end, will depend on general principles of law. A short definition of a contract is 'an agreement between two or more parties which is intended to have legal consequences'. In construction, the contract is generally for producing a building or part of the built environment, and can be entered in one of four ways:

1. Implied by conduct of the parties; a contractor may submit an offer and later have access to the site;
2. By word of mouth; typically where an offer is accepted by telephone;
3. By exchange of letters; common for small domestic works of alteration or repair;
4. Using a written contract; the contract documents often include the enquiry documents, the written offer, minutes of meetings, a programme, a method statement and a formal contract with the agreed terms.

An estimator should keep a separate file containing all papers which will form the basis of the agreement. This is most important where negotiations take place after a formal offer has been made. If the estimator secures the work, he will need to present the contractor's undertakings to the construction staff at a handover meeting. The importance of written evidence cannot be overstressed, because usually the formal documents will be the only evidence of what exactly had been agreed at the beginning of a project.

To make a contract valid and legally enforceable, certain simple rules are applied, as follows:

1. There must be an offer by one party and an acceptance by the other or others.
2. Each party must contribute something of value to the other's promise; a client is responsible for making payments and the contractor must complete the construction.
3. Each party must have the legal capacity to make a contract.
4. The parties must have exercised their own free will, without force or pressure.

A contract comes into existence when an offer has been unconditionally accepted. In construction the offer is the 'tender', 'estimate' or 'bid' and suppliers and sub-contractors sometimes refer to their offers as 'quotations'. The term 'estimate' could be used in a wider context to mean a guide to how much something will cost. This ambiguity should be avoided wherever possible.

A contractor expects to receive an acceptance in clear terms from the client or his adviser. A letter of intent is often used to let a contractor know that he should prepare to start work. This statement should state clearly that all work carried out by the contractor and specialists, even if the contract does not follow, will be paid for in full.

An offer must be distinguished from an 'invitation to treat' which is an invitation for others to make an offer. In an auction sale, for example, an auctioneer invites offers which he may accept or reject. In a similar way, a client seeking tenders is not bound to accept the lowest or any bid. An offer cannot be accepted once it has terminated. Termination happens:

1. On the death of either party if the contract is for personal services;
2. By the contractor withdrawing the offer;
3. After a specified time (usually stated in the tender instructions or by the contractor in his tender) or after a reasonable time;
4. When there has been outright rejection by the client, or where the client makes a counter-offer, usually in the form of a qualified acceptance.

Although contractors and sub-contractors can withdraw their tenders at any time before acceptance, this practice can lead to many problems for the recipient. A main contractor, awarded a contract, could lose a large sum of money if a sub-contractor's offer, used in a tender, is withdrawn or changed. The main contractor should clearly state in his enquiry documents the acceptance period for sub-contractors' tenders taking into account the requirements of the main contract and the possible delay in placing contracts.

Standard forms of contract

The standard printed forms of contract have been developed over many years to take account of the many events which could occur during and after a construction project. Contract law will, of course, deal will many of the problems, but there are many matters peculiar to construction which need clarification. Once these terms have been incorporated, they reduce the likelihood of disputes which can lead to arbitration or litigation. Contract conditions are outlined by a reference being made to the standard conditions in the tender documents, with amendments to suit the particular project. The parties to most of the JCT contracts sign copies of the printed forms, which is not the case for the ICE and GC/Works/1 forms, which could be used by reference to an 'office' copy. Some clients require a contract to be executed under seal; the standard forms have provision for this after the Articles of Agreement. A contract executed as a deed (or speciality contract) would allow an action to be brought within twelve years as opposed to six years for simple contracts. It is unwise to amend the conditions of a standard form because great effort has gone into producing a carefully drafted document with many links between clauses and other documents. Nevertheless, all contracts take effect by agreement and so standard contracts can be amended in any way the parties choose.

The standard form contracts currently in use between client and contractor are:

1. JCT Standard Form of Building Contract   JCT80
2. JCT Standard Form with Contractor's Design   CD81
3. JCT Agreement for Minor Building Works   MW80

4. JCT Intermediate Form of Building Contract   IFC84
5. JCT Standard Form of Management Contract   MANCON87
6. GC/Works/1 Ed 3 for Government Contracts   1989 rev90
7. ICE Conditions of Contract 6th Edition   1991
8. ICE Design and Construct Conditions of Contract   1992

The Standard Form of Building Contract, written by the Joint Contracts Tribunal, is made up of bodies representing differing interests in building work, including the RIBA, BEC, RICS, Local Government Associations, consulting engineers and specialist contractors' associations. The standard form has six variants which cater for local authority and private clients, contracts with bills of quantities, without quantities and those with approximate quantities. The six forms do not differ in substance, but describing and costing the work is easier with bills of quantities. The local authority forms are similar to private forms but contain extra terms for local government law and practice. Each variant creates a lump-sum contract: the lump sum is that which the contractor expects to be paid but is subject to adjustment in many carefully defined ways, mainly following the issue of an instruction. A bill of quantities is also used with the ICE conditions for civil engineering work, but the conditions create a remeasurement or 'measure-and-value' contract where all the bill items will be remeasured as the work proceeds. The ICE conditions are alone in defining permanent and temporary works; the ICE form makes it clear that temporary works are solely the responsibility of the contractor except where they have been designed by the Engineer. The GC/Works/1 contract is used for building and civil engineering works.

The appendix section of standard forms enables the parties to insert provisions which vary from job to job, such as:

1. *Sums of money* for liquidated damages and insurances
2. *Periods of time* for carrying out the work and making payments
3. *Percentages* for retaining parts of the interim payments
4. *Statements* giving the options which apply to the contract. An important example would be to show which clause has been selected for dealing with price fluctuations.

This information must be given to tenderers, otherwise they will make their own assumptions.

Sub-contract forms

The contractual links between parties using standard forms of contract are shown in Figs 3.1 and 3.2.

Nominated sub-contractors are persons whose final selection and approval, for supplying and fixing materials or goods, has been reserved to the architect (Clause 35 JCT80). In the ICE conditions nominated sub-contractors are 'any merchant tradesman specialist or other person firm or company nominated in accordance with the contract to be employed by the contractor for the execution of work or supply of materials for which a prime cost has been inserted in the contract . . .'. There is usually a right of objection to nominating a sub-contractor because it would be contrary to contract law to insist that a party enters a contract involuntarily.

Domestic sub-contractors are engaged where a contractor elects to sub-let part of the work with the written consent of the architect. (Clause 19 JCT80). The ICE contract goes on to say that the contractor shall not sub-let all the works without the written consent of the Employer.

'*Named sub-contractor*' is the term used in IFC84 and ACA84 where the contractor is required to enter a (domestic) sub-contract with a firm named by the architect.

Many contractors sub-let large portions of their work to specialist contractors, the main exceptions being where reliable building workers are needed for difficult or small maintenance contracts. Under clause 19 of JCT80, there are two arrangements for sub-letting work to domestic sub-contractors:

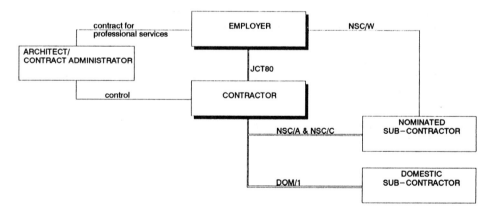

Fig 3.1 *Contractual relationships between parties using JCT80*

25

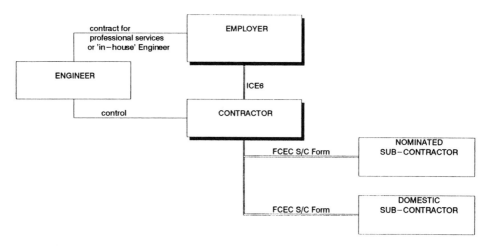

Fig 3.2 *Contractual relationships between parties using ICE 6th Edition*

1. The architect approves the sub-letting of the works to a firm of the contractor's choosing.
2. The contractor must choose a sub-contractor from a list of at least three names which have been included in the contract bills, for work fully measured in the bills and priced by the contractor.

The latter arrangement is used sometimes to replace nominated sub-contractors with a shortlist of specialists who may have expressed an interest in doing the work. Where large service installations are required, the quantity surveyor can send the drawings and specification to each sub-contractor on the list so the main contractors can avoid unnecessary duplication. The estimator just sends his enquiry letter with details of the conditions which the sub-contractor will be expected to sign.

If a single firm is named in the contract bills to carry out work which is measured, then it should in effect be a nominated sub-contractor. In fact this would be bad practice because SMM7 requires a PC sum for work to be executed by a nominated sub-contractor. This problem is unfortunately all too common. An example is where a client or consultant wants a particular window system but wishes to avoid setting up a formal nomination. The standard form of contract does not allow the architect to choose a specialist who is to become a domestic sub-contractor.

The standard forms commonly used between contractors and their sub-contractors are:

1. DOM/1: Domestic sub-contract approved by BEC, FASS and CASEC for use with the JCT80 main forms, with or without quantities. These conditions are widely used by main contractors often with amendments.
2. DOM/2: Domestic sub-contract for use with the JCT Standard Form With Contractor's Design 1981. The document comprises the articles of agreement with a schedule of changes, such as:
   (a) Delete 'Architect' and insert 'Employer';
   (b) Provision for stage OR periodic payments.
3. FCEC: Form of Sub-contract for use with the ICE Conditions of Contract (commonly called the 'Blue Form').
4. NAM/SC: Conditions for named sub-contractors under the Intermediate Form of Contract 1984.
5. IN/SC: Domestic sub-contract for use with IFC 1984.
6. GW/S: Standard form of sub-contract for use with GC/Works/1; published by BEC and approved by CASEC and FASS. The terms of this sub-contract stem from DOM/1, NSC/4 and obligations passed down from GC/Works/1.

Most sub-contract forms are printed in two parts: the articles of agreement and conditions. This could be to save money since only the articles of agreement are needed each time contracts are signed.

Non-standard forms of sub-contract are sometimes used by main and management contractors to impose extra obligations and ensure the sub-contractor is bound by the same conditions found in the main contract. The trade bodies which represent the views of specialist sub-contractors claim that their members are suffering under terms such as:

1. The 'pay when paid' arrangement which means that a sub-contractor will be paid when the main contractor has received a payment.
2. The 'discount fiddle' happens when the 2.5% discount for prompt payment is held by the main contractor well beyond the agreed time.
3. Reduced attendances provided by main contractors, in some cases expecting sub-contractors to provide their own scaffolding, temporary services, disposal of rubbish and hoisting.
4. The sub-contractor's right to an extension of time might only be granted when the main contractor himself receives an extension.
5. The main contractor can hold wide-ranging rights to take sums of money from payments, sometimes without having to prove that a loss has occurred.
6. A requirement for a sub-contractor to protect his work even when he is not present on site.
7. Badly drafted 'on demand' bonds and parent company guarantees irrespective of the size or stature of the company.

It is becoming more common for main contractors to be on the receiving end of some of these practices. In particular, some clients want set-off clauses and performance guarantees which can be taken 'on demand' and may be kept in place for a long time after the project is complete. Both main and sub-contractors when faced with such enquiries should return their tenders with a statement asking to discuss the terms of contract with the client before entering a formal agreement.

The practice of nominating sub-contractors is on the decline because although the main contractor is contractually responsible for all the works there is a reduced liability for the work sub-let under the nomination system. The JCT80 Form of Contract makes the following provisions:

1. Delay by a nominated sub-contractor is a relevant event which can lead to an extension of time under clause 25.
2. Breach by the nominated sub-contractor imposes a duty on the architect to nominate a new sub-contractor if the first is incapable of performance.
3. Failure of design by a nominated sub-contractor under clause 35.
4. Delay caused by a nominated sub-contractor who gives late information.

In a traditional contract, where a decision to adopt a particular supplier or sub-contractor is needed before appointing a main contractor, a nomination is required. It allows the architect to prepare full working drawings, integrating and coordinating specialist design with building design. Costs are saved at tender stage because the specialist prepares one tender on a standard set of conditions.

The main documents for nomination introduced by the JCT80 contract are:

1. NSC/1: JCT Standard Form of Nominated Sub-contract Tender and Agreement is used to invite tenders from potential nominated sub-contractors which gives the sub-contractor information about the contract, allows the sub-contractor to submit his tender and later agree programme and attendance details with the main contractor.
2. NSC/2: JCT Standard Form of Employer/Nominated Sub-contractor Agreement which sets out the obligations of the sub-contractor to exercise reasonable skill and care in the design of the works and perform satisfactorily when under contract to the main contractor. It is important to remember that neither the architect nor the contractor are party to this contract which is used to provide a warranty agreement to protect the client's interests. The client also has obligations, mainly to pay for design work and materials before and after the start of construction works.
3. NSC/3: JCT Standard Form of Nomination is an instruction from the architect for the contractor to enter a sub-contract.
4. NSC/4: JCT Standard Form of Nominated Sub-contract which should be read in conjunction with clause 35 of the main contract form.

28

In publishing the tenth amendment to JCT80 the Joint Contracts Tribunal has introduced a new series of documents which are intended to improve the complex arrangements in use since 1980. Fig. 3.3 shows the old arrangement of four documents which have been replaced by seven. The changes are cosmetic and the contractual relationships have not been affected.

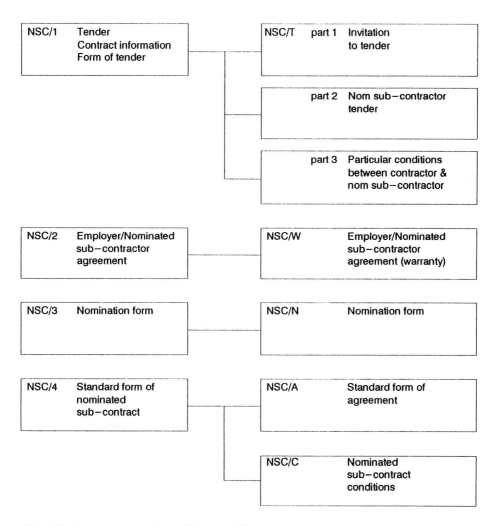

Fig 3.3 *A comparison of the 1980 and 1991 nomination documentation*

Comparison of forms

Contract documents all include the various parts of the standard forms plus:

JCT80	Contract drawings, contract bills
CD81	Employer's requirements, contractor's proposals (including contract sum analysis)
MW80	Contract drawings, contract specification, priced specification or schedule of rates
IFC84	Contract drawings, contract specification, schedules of work or bill of quantities
MANCON87	Project drawings, the project specification, Contract cost plan and the schedules
GC/Works/1	Drawings, specification, bills of quantities or schedule of rates
ICE	Drawings, specification, bill of quantities, tender and the written acceptance
ICE Design & Construct	Employer's requirements, contractor's submission and the written acceptance together with other agreed documents.

The employer may deduct a *retention* as follows:

JCT80	5% until practical completion then 2.5% (3% if estimated value is over £500 000)
CD81	5% until practical completion then 2.5%
MW80	5% until practical completion then 2.5%
IFC84	5% until practical completion then 2.5% (no retention on insurance payments)
MANCON87	3% for management contractor and works contractor (not applied to fees) and 1.5% for work which has reached practical completion
GC/Works/1	5% of each advance payment and nil for the value of variation instructions
ICE	Recommended not to exceed 5%, recommended limit is 3% of tender total, and retention halved following certificate of completion.

All the standard forms contain terms for insurances but only the ICE conditions include a recommended form of bond. The new (Edition 3) of the GC/Works/1 contract has brought in some radical changes which have an effect on the tender. For example, there is no longer a discount retained by the Contractor from PC sums for nominated sub-contractors. Valuations made monthly by reference to a pre-determined stage payment chart would undervalue the work at the beginning of a contract. The Minor Works agreement does not cater for long projects which need a price fluctuations clause and does not provide for nominations by the supervising officer.

Selection of contract forms

For many clients the choice of contract will be dictated by the type of work, size of contract and their position in society. A local authority carrying out a £2 million

CONTRACT SELECTION CHECKLIST

Procurement method	Lump sum	
	Measurement	
	Cost reimbursement	

Cost control document	Bills of quantities	
	Schedule of rates	
	Priced specification	
	Contract sum analysis	

Payment	Stage	
	Time–related	
	Turnkey	

Roles and relationships	Client	
	Contractor	
	Design team	
	Specialists	

Time	Open	
	Fixed	
	Acceleration	
	Damages	

Fig 3.4 *Simplified checklist for the selection of a contract*

31

refurbishment contract, for example, is likely to choose the JCT Standard Form of Building Contract, Local Authorities Edition, with approximate quantities. A contractor offering his services to design and build a factory unit will suggest the Standard Form with Contractor's Design CD81. Perhaps the most difficult decisions to be made by a client are the composition of the professional team and how financial risks will be shared. In particular, he must decide whether to commission a bill of quantities or ask for tenders on a lump sum. Figure 3.4 shows the primary elements which need to be considered. Clearly a non-construction client would need professional advice in selecting a contract which satisfies all his needs.

Tender documentation

Introduction

The key to a successful project often lies in the understanding and cooperation that is essential from all participants; each must be clearly aware of his duties and rights. The documentation is the vital link between design and construction.

Adequate and accurate drawings and specifications are indispensable if the team is going to achieve success in terms of quality, time and cost. Drawings in particular have served the construction industry well for hundreds of years as the primary means of communication. Unfortunately, poor specification writing continues to be a weak link in the information chain and leads to disputes, particularly in a competitive market where estimators will use a strict interpretation of the documents to arrive at the lowest tender. Another cause of friction is bills of quantities which differ from the drawings and specification. This often happens when the quantity surveyor is short of information from the designers.

Time spent on preparing documents, which aid the contractor's understanding of the work, will benefit the finished product. In 1964 the report of the committee chaired by Sir Harold Banwell stated:

> It is natural that a client, having taken the decision to build, should wish to see work started on site at the earliest possible moment. It is the duty of those who advise him to make it clear that time spent beforehand in settling the details of the work required and in preparing a timetable of operations . . . is essential if value for money is to be assured and disputes leading to claims avoided. It is also necessary for the client to be told of the need to give the contractor time to make his own detailed arrangements after the contract has been let, and of the penalties of indecision and the costs of changes of mind once the final plans have been agreed.

Tenderers will assess the quality of documentation, partly because poor information can add to the time wasted by site supervisors and partly because unreliable information can lead to claims. If the contractor has enough information he can avoid guesswork, include all the important items in his tender and will not need to add global sums for poorly defined elements of work.

Coordinated project information

The Coordinating Committee for Project Information was set up in 1979 to look for improvements in the way construction documents are produced and presented. The committee published its recommendations in December 1987 for drawings, specifications and bills of quantities for building work; and included proposals for ways in which the following problems may be overcome:

1. Missing information – not produced, or not sent to site
2. Late information – not available in time to plan the work or order the materials
3. Wrong information – errors of description, reference or dimension; out-of-date information
4. Insufficient detail – both for tender and construction drawings
5. Impracticable designs – difficult to construct
6. Inappropriate information – not relevant or suitable for its purpose
7. Unclear information – because of poor drafting or ambiguity
8. Not firm – provisional information often indistinguishable from firm information
9. Poorly arranged information – poor and inconsistent structure, unclear titling
10. Uncoordinated information – difficult to read one document with another
11. Conflicting information – documents which disagree with each other.

Drawings

Drawings are the most common means of communication for all types and sizes of project; the main exceptions being some maintenance contracts and minor works which can be scheduled or described in a written statement. The CPI initiative includes a production drawings code which gives advice on good practice for planning and producing drawings. The code stresses the need for careful coordination of the information, shown on drawings, with the other documents. One way to avoid mistakes is to replace specifications on drawings with reference numbers which refer to the written specification. This could, however, lead to confusion on site if taken to an extreme case such as a drainlayer asked to lay a drain R12/123 in a trench type R12/321. Would he need to be armed with the drawing and specification? Probably not; because designers understand the need for clear information for those working on site and on large-scale projects site engineers interpret the drawings for the operatives.

The CPI code is to be read with BS 1192: 1984 'Construction Drawing Practice'. This British Standard was rewritten during the 1980s and published in four parts. This revision has been brought about by the need for international standardization of drawing practice; and many industrialized countries have taken part in the search for suitable conventions and methods. The aim of the new standard is to provide good drawing practice which will provide communication with:

1. Accuracy
2. Clarity
3. Economy
4. Consistency

between architects, contractors, civil engineers, service engineers and structural engineers.

There are four main types of drawings commonly used in construction:

1. Survey drawings – which are based on a measured survey or an Ordnance Survey sheet; and are used to produce block and site plans.
2. Preliminary drawings – which are the designer's early interpretation of the brief.
3. Production drawings – include general arrangement drawings, layout drawings, assembly drawings, standard details such as those provided for highways drainage, schedules and additional detail drawings as necessary. They are used to go with applications for statutory approvals, to invite contractors to tender, and construction purposes.
4. Record drawings – are used to show a record of construction as it has been built and services installed. They provide essential information for maintenance staff.

Since the publication of SMM6 some drawn information can now be provided with bills of quantities. SMM6 recommended the use of bill diagrams to help when describing an item of work. For example, rule F.15.5 dealt with formwork to beams and projecting eaves, given in metres. Where a member is inadequately described in this way, each description is to be accompanied by a bill diagram showing the required profile.

In SMM7, general rule 5.3 states 'dimensioned diagrams shall show the shape and dimensions of the work covered by an item and may be used in a bill of quantities in place of a dimensioned description, but not in place of an item otherwise required to be measured'. The intention is for these diagrams to be prepared by the quantity surveyor and included in the bill of quantities. Often this has not happened with either SMM6 or SMM7. This might be because bills are produced using text-based computer systems and more drawings are now sent to contractors at tender stage.

Specifications

A specification is prepared by an architect or consulting engineer to provide written technical information mainly on the quality of materials and workmanship. The specification would be a contract document in its own right if the contractor tenders on the basis of drawings and specification only. Where bills of quantities are used for building work the specification is included with the bill of quantities as preambles. In this way the specification again becomes part of the contract documents.

There are some standard specifications published for civil engineering contracts, particularly specifications for highways and the water industry. A bill of quantities for civil engineering work will include specification clauses and a preambles section which is used to define the method of measurement and any departures from the standard method.

The designer notes the matters needing detailed specification clauses as he prepares the drawings. The quantity surveyor will advise on a proper format for the bill of quantities. On small contracts, where a PQS is not appointed, an architect could produce a specification which is broken down into parcels of work. The contractor would be expected to price the document to assist post-contract cost control, such as the preparation of valuations. In this context, this document is sometimes called a schedule of works.

The CPI code for specification writing is a guide to good practice. Many architects, engineers, quantity surveyors and contractors will subscribe to the National Building Specification, which has been rewritten in line with the new Common Arrangement of Work Sections (CAWS). The National Building Specification is a library of clauses, regularly updated, using either the CI/SfB classification or the recommended CAWS method which divides building into over 300 work sections which aim to reflect the way work is sub-contracted. In broad terms CI/SfB relates to the elements of a building and the CAWS is in trade order. Normally, only a fraction of the work sections will be used on a simple project.

Specifications are prepared by design teams (or contractors in the design and build contract) using their own procedures and often vary widely in coverage and technical content. It is often said that specifications have lagged furthest behind drawings and bills for quality and helpfulness.

There is a danger that specifications may be ignored by contractors, sub-contractors and suppliers because they:

1. Contain many standard clauses which are not relevant to the job
2. Are usually too long
3. May be a collection of protection clauses, for example: 'to the best quality', because the designer is not sure what quality to specify
4. Are frequently out of date.

Traditionally the architect has been responsible for the specification, but may delegate the printing to the PQS. The CPI initiative assumes that more reliable specification information is provided before tender stage by the designer. The PQS must ensure the bill descriptions do not conflict with the specification. With the introduction of SMM7, bill descriptions will include cross-references to the specification which will remove duplication.

The Project Specification Code recommends improvements, so specifications will be:

1. Complete – covering every significant aspect of the work
2. Project-specific – produced for the project, without irrelevant material
3. Appropriate – for available materials and skills; and can be checked and standards enforced
4. Constructive – helping all the parties to understand what is expected of them
5. Up-to-date – using current good building practice and most recent standards
6. Clear – economically worded.

Bills of quantities

The traditional purpose of bills of quantities is to act as a uniform basis for inviting competitive tenders, and to assist in valuing completed work. Bills of quantity are first designed to meet the needs of estimators, although some estimators say that the bill format has changed to assist the consultants in cost planning exercises through the widespread use of elemental bills.

A contractor can also make use of the bill of quantities in many ways:

1. To plan material purchasing (note the danger in ordering from a bill: the contractor should always order materials from drawn information and the specification, making the contract administrator aware of any differences);
2. Preparing resourced programmes;
3. Cost control during the contract to ensure work is within budget;
4. Data collection during construction for bonus systems and feedback information for estimators.

Unlike drawings and specifications, there have been rules for measuring building work for many years. The first edition of the Standard Method of

Measurement for Building Works was published in 1922 and has been a compulsory document since its incorporation into the RIBA (now JCT) contract 1933. The civil engineering methods include rules for highways and the water industry but the publication for mainstream civil engineering works is the Civil Engineering Standard Method of Measurement (CESMM3) now in its third edition (1991).

Bills of quantities for building are divided into the following sections:

1. Preliminaries
2. Preambles
3. Measured work
4. Prime cost and provisional sums.

There are number of formats for Civil engineering bills of quantities. CESMM3 gives the following sections:

1. List of principal quantities
2. Preamble
3. Daywork schedule
4. Work items (Class A General items may be grouped in a separate part of the bill of quantities).

In both cases, the estimator prices sections 3 and 4 and the specific items described in the preliminaries, having taken full account of all the requirements in the other sections.

The preliminaries (general items) section gives general details about the project and contract conditions, as follows:

1. Description of the work, location of the site, site boundaries, names of parties, and lists of drawings;
2. The form of contract used, with any amendments clearly defined, with contract appendix details giving information such as the retention percentage, liquidated damages, possession and completion dates and fluctuation provisions;
3. Specific requirements which should be priced by the contractor as fixed or time-related items to reflect the actual costs arising from supervision, site accommodation, temporary works, site-running costs, general plant, transport, client's requirements and safety.

CESMM3 and SMM7 provide for fixed and time-related items so that a contractor can show the cost of bringing plant or facilities to site, their maintenance during the job and removal on completion. The SMM7 Measurement Code suggests that work will be priced in this way only if the tenderer wishes to do so. There should also be space in the preliminaries section of a bill for the contractor to add to the list of items to suit his particular methods of working. In CESMM these are called 'method-related charges'.

In bills of quantities for building work the preambles contain specification clauses which provide information about the expected type and standard of materials and workmanship. They should relate to the work in the bill and will increasingly reduce the work description.

The measured work section of the bill of quantities is divided into trade or element headings and measured according to the rules of a standard method of measurement. SMM7 defines its role by the statement 'The standard method of measurement provides a uniform basis for measuring building works, and embodies the essentials of good practice. Bills of quantities shall fully describe and accurately represent the quantity and quality of the works to be carried out'. The Standard Form of Building Contract JCT80 requires the use of the standard method of measurement where the contract includes bills of quantities. Clause 2.2.2.1 states 'the contract bills shall be prepared in accordance with SMM7'.

Accuracy in preparing a bill is essential because the contract conditions allow the contractor payment for any omission or error in description or quantity. Clause 2.2.2.2 states '. . . an error is treated as a variation' and clause 8.1 states '. . . all materials, goods and workmanship shall be to the standards described in the contract bills'. Clause 1.10.3 JCT Work Contract/2 used with the Management Contract similarly states 'the quality of the work included in the Works Contract Sum or Tender Sum shall be deemed to be that which is set out in the bills of quantities'.

SMM7 begins with general rules for preparing bills, followed by details of preliminary particulars and about 300 work sections under 24 main headings. Rule 4.1 is an example of a rule of particular interest to an estimator:

> Dimensions shall be stated in descriptions generally in the sequence length, width, height (or depth). Where ambiguity could arise the dimension shall be identified.

Where work can be identified and described in a bill of quantities, but the quantity cannot be accurately determined, an estimate of the quantity can be given and identified as an 'approximate quantity'. This will typically occur when dealing with ground problems such as stone filling under a ground beam or maintenance work such as cutting out defective rafters.

A client provides a provisional sum in a bill of quantities for work which cannot be described and given in items, which follow the measurement rules. SMM7 introduced two kinds of provisional sum, defined and undefined, both for work which is not completely designed. 'Defined' means that the nature and quantity of the work can be identified, and the contractor must allow any planning and overhead costs in his calculation of project overheads. 'Undefined' means that the scope of the work is not known, and the contractor will be paid for all costs associated with carrying out the work, planning the work, and overheads which are reasonable.

A contingency sum is often included in a bill, as a provisional sum, for unforeseeable work, such as difficult ground conditions. The reason for its inclusion is not stated in the bill. The sum is spent at the discretion of the architect/contract administrator. SMM7 does not mention the contingency sum.

A prime cost sum is provided in a bill of quantities for work to be carried out by a nominated sub-contractor (SMM7 A51) or for materials to be obtained from a nominated supplier (SMM7 A52). Work by statutory authorities is now given as a provisional sum (SMM7 A53). SMM7 does not define PC sums to the extent found in SMM6 presumably because the form of contract deals with this. The term 'prime cost' is also used in connection with:

1. An allowance for the cost of a material such as bricks when the final selection has not been made; for example, Facing Brickwork PC £150.00 a thousand (the estimator must be told how to deal with waste, transport and other on costs).
2. The basic cost of labour, materials and plant in cost-plus arrangements such as daywork contracts and some management contracts. SMM7 now gives dayworks as a provisional sum (A55).

The SMM7 Measurement Code recommends clear information about nominated sub-contractors' work, so that the tenderers can assess their responsibilities, for their programme. At tender stage the contractor should know:

1. The extent of the nominated sub-contractor's work with approximate quantities or values in each part of their work;
2. The location of the work, in particular where large items of plant will be situated;
3. Special attendance which is needed by the specialists, with the location and dimensions being given wherever possible (if details of special attendance are not available then a provisional sum should be used).

SMM7 provides for certain drawings to be issued to contractors at tender stage. More detailed guidance on which drawings are needed is given at the beginning of each work section in SMM7. The following drawings are considered to be essential:

1. Block plan
2. Site plan
3. Plans, sections and elevations.

Component drawings are required by general rule 5.2 to show the information necessary for the manufacture of components. The work sections which require component drawings are listed in Appendix 2 of the Measurement Code.

Bill formats

The development unit, which prepared SMM7, made some general recommendations for good practice, as follows, and included some of them in the SMM7 Measurement Code:

1. The full benefits of the CPI initiative will be gained if bills and specifications are prepared using the Common arrangement of work sections (CAWS).
2. Items for separate buildings should be kept separate, by providing separate bills.
3. Items for external works should be given in a separate bill.
4. Provisional sums, prime cost sums and dayworks should form a separate section at the end of the measured work part of the bill (avoiding confusion during the tender stage).
5. The summary should be at the end of the bills of quantities.

Estimators have a strong preference for trade bills which separate work strictly in accordance with the measurement rules and trade headings of SMM7. This is convenient for sending enquiries to suppliers and sub-contractors.

Elemental bills relate to the functional parts (or elements) of a building; for example, roofs, external walls or upper floors. This has the benefit of helping the quantity surveyor check his cost analysis and collect data for future cost exercises, and the location work can be found by the estimator. The main disadvantage is that it produces a longer bill which is not only less efficient to prepare but will also add to the work of the estimator. He must bring together items for each trade from various parts of the document which can produce a great deal of paperwork.

Sectionalized trade bills are an attempt to overcome the disadvantages of the elemental bill. For estimating purposes the trade order bill is subdivided into elements. If each element is printed on separate sheets, it is possible to assemble the bill in trade *or* elemental order.

Computer packages are available for producing bills of quantities and subsequent financial control. They are usually based on a library of standard items which can be called up by using codes or by accessing a hierarchal database through menus. Measurements can be entered either manually or using digitizers and the computer will sort the items before printing the complete bill of quantities.

CPI and the estimator

The common arrangement of work sections was developed to align packages of work more closely with the pattern of sub-contracting in the industry. For example, SMM7 now clearly distinguishes between many cladding methods and materials in group heading H, Patent glazing, Curtain walling and many kinds of sheet cladding. Unfortunately, this fine subdivision has some awkward results. For example, an enquiry for plumbing will include measured work from Group R Disposal systems, Group S Piped water supply systems, Part T Mechanical heating, Group N Sanitary appliances and Group Y for pumps and calorifiers. Furthermore, with an elemental bill any of the 300 work categories can be repeated for each element.

Two of the objectives of SMM7 were: (1) to simplify bills of quantities and (2) to develop a method which could help with computer applications. Since the building industry has been slow to adopt the CPI conventions, it is too early to measure its success.

It is not yet clear whether SMM7 will produce shorter bills of quantities. The theory was that many items could be removed where they had little cost significance. Other items could be grouped, again to lessen the number of measurable items of work. The nominal size of bar reinforcement is stated but its location is not. This means that the bill rate for 12 mm reinforcement in an eaves beam will be the same as 12 mm bars in a ground-floor slab. An estimator may be able to identify the weights in each location by studying the drawings and bar schedules, but will all the estimators and sub-contractors do the same? Since SMM6 was introduced, estimators have been faced with formwork measurements grouped in height bands which effectively mean that they do not know the actual quantity. As an example, 200 linear metres of formwork 500 mm to 1.00 m high could be as little as 100 m² or as much as 200 m² of shuttering. It would take some time to measure the real area from the drawings received at tender stage.

The move towards computer-aided billing and estimating has been difficult, and many doubt whether SMM7 has helped. On first sight, the tables of measurement rules appear to be an aid to all those involved in computer-aided bill production and pricing. Unfortunately there have been some problems:

1. There are too many rogue items in an average bill which do not match a standard coding system.
2. The PSA library of standard descriptions does not use the numbering given in SMM7.
3. Computer packages have moved away from code numbers for items, preferring to use windows of items from which relevant descriptions and resources can be selected (code numbers are still used for resources because it

is essential to collect similar resources to complete the summaries for the adjudication meeting).

It is also argued that the standard method of measurement is not one for producing bills nor is it an aid to pricing bills. It is purely a set of rules about how work is measured and what is to be included in an item (item coverage).

The civil engineering standard method of measurement CESMM3 states that the system of work classification adopted by the method should simplify the production of bills of quantities, making the use of computers easier. The foreword to the First Edition encourages the use of work reference numbers to identify work items. This uniform (and coded) description of work was seen as a way to standardize on the layout and contents of bills of quantities; and the engineer is recommended to use the standard method numbers in bills of quantities. This recommendation does not exist in SMM7.

Every estimator, whether working for a contractor or sub-contractor, must understand the coverage rules of the standard method which applies to the contract. This is important where sundry items are now included in the main work item. For example, in SMM7 formed joints in *in-situ* concrete are deemed to include formwork; and working space allowance must include the extra cost of work below groundwater level and breaking out existing hard materials.

By making the detailed specification the central reference document under CPI, the way in which estimators work has changed. Enquiries to sub-contractors and suppliers must include all relevant specification clauses, preliminary section items and appropriate drawings; otherwise the prices will not reflect the true value of the work. Bill descriptions will become shorter by constant references to the specification. The following example illustrates the problem:

Forming cavities in hollow walls
60 mm wide; wall ties spec 310;
cavity insulation spec 560, 30 mm
thick,   115 m².

If a sub-contractor receives an incomplete enquiry, he is likely to guess what he is being asked to fix.

Experience of pricing documents, which have been produced using the CPI guidelines, shows that some new problems have emerged, as follows:

1. Many specifications have no page numbers. The explanation is that the estimator must use the NBS codes to find the relevant clauses. The problem for the estimator is that the page numbers are needed for printing and distributing pages to the sub-contractor.
2. There is confusion with the way work section numbers are used in

specifications. On one page the estimator might find clause 310 which is for laying bricks and on another page clause 310 could be for cavity wall ties. The problem is that the work section reference is missing, in the first case it should be F10.310 and the second F30.310. The work category numbers must be repeated on each new page if this problem is to be solved.

3. Some sub-contractors have argued that they received the bill but not the specification. With SMM6, the bill description often had enough detail to price the work. This may be the estimator's fault but in some cases the specification references have become complex. The estimator might find the correct clause referred to in the bill but not notice that the specification clause includes references to other clauses.

4. Defined provisional sums are being used incorrectly. They should provide information about the nature of the work, a statement about how and where the work is fixed, quantities to show the scope of work, and any limitations. It is common to see defined provisional sums such as: 'drainage outfall to culvert' or 'additional dry-rot treatment'.

5. The number of drawings needed by sub-contractors at tender stage has increased considerably. This is due to the reduction in the number of bill items; or as some would say, 'the quantity surveyor doing less work'.

Now that main contract bids rely heavily on quotations from sub-contractors, the estimator must exercise great skill and care in dealing with changing procedures and new methods of measurement for shorter bills of quantities. The PQS still has the responsibility to provide adequate information for the estimator to price. As SMM7 insists, 'More detailed information than is required by these rules shall be given where necessary in order to define the precise nature and extent of the required work'.

Documents used as the basis of a tender

The basis of the tender will dictate the way in which the contractor will be paid and the relative accuracy of the estimate. The contractor's bid will be for one of the following:

1. Fixed-price contract: where the sum of money is stated in the contract as payment for work, the payment may be adjusted according to strict conditions in the contract.

2. Measurement contract: will allow the contract sum to be calculated later, usually as the aggregate of various rates submitted by the contractor. The contract sometimes includes a target price.

3. Cost-reimbursement contract: an arrangement by which the cost, whatever it may be, will be paid by the client on the basis of the actual cost incurred by the contractor.

Fixed-price contracts

The price is fixed in advance but is subject to variation under the terms of the contract. This could include a fluctuations clause to pay for the increases caused by inflation. This definition leads to much confusion in the construction industry where 'fixed price' is the term for a price which will not be subject to fluctuations. An arrangement which is not subject to fluctuations is better described as firm price.

Lump-sum contracts are the simplest type where a lump-sum offer is made by a contractor to carry out the work, which might be outlined on drawings and described in a specification but no quantities have been prepared. This is the usual form for a small job carried out by a local builder. Where a full set of working drawings and a specification are available, a drawings and specification ('plan and spec') arrangement is popular for small projects. The main advantage is the saving in time and money needed to prepare a bill of quantities. The client will also have a reasonable estimate of the total cost before the contract is signed. For small contracts where the client's requirements are clear and there are good drawings and specification, this can be a useful way to enter a contract. There are, however, some serious drawbacks. Each contractor must prepare his own bill of quantities and the employer must bear in mind the time needed during the tender stage. If construction details or specification requirements are missing, it is common to find each contractor tendering on different assumptions. The contractor must allow a contingency for the risk of making mistakes in taking-off. There will be no detailed breakdown of the tender sum which would be needed for interim payments and for valuing variations. To overcome some of these disadvantages, the specification should include a description of work in a series of numbered items, each of which is to be priced.

Bills of quantities provide the most detailed basis for estimating cost. Each contractor tendering for work will be familiar with their use and can save wasteful effort in preparing quantities for the same building. There will be a clear list of items included in the contract and a schedule on which variations may be valued. They give a fair basis for competition, and a firm contract sum is known in advance. The main disadvantages are the time needed for the accurate preparation of bills (less of a problem with computer techniques) and the bill rates are not able to deal with variations which change the character of the work. Firm bills of quantities remove the onus for correct quantities from the contractor but may inflict higher charges on the employer if discrepancies exist between the documents.

When a contractor tenders for a design and build project, he prepares his own bill of quantities (from his own drawings and specification) in order to invite sub-contract bids and arrive at a cost for direct work. The contractor's proposals include a contract sum analysis. The purposes of the contract sum analysis are:

1. To value changes in the employer's requirements;
2. To value provisional sums given in the employer's requirements;
3. To allow the use of price adjustment formulae where they apply.

The contract sum analysis should be divided into sums of money for design work carried out before and during construction, and the following:

1. Preliminaries
2. Provisional sums
3. Trade headings similar to those in SMM6 or SMM7.

Work in different buildings and external works is usually shown separately.

Bills of approximate quantities provide a fair basis for tendering when drawing details are not complete. The bills will represent an estimate of the quantities of work in the project. By definition, the work will be subject to remeasurement and a firm value will not be known at the start of the project. This method is commonly used with refurbishment work where the full extent of the work cannot be accurately determined. The quantities set out in bills of quantities for civil engineering are the estimated quantities and are not to be taken as the actual quantities; the actual quantities are measured during the construction phase.

Measured contracts

The total cost of a contract can be calculated by measuring the work as it advances on site and pricing the measured items using the rates given in an agreed schedule of rates or approximate bill of quantities. A schedule of rates lists all the items likely to arise, in a similar way to a bill of quantities, but no quantities are included. A schedule of rates is also used with drawing and specification contracts to value additional work. There are two principal types of schedule.

A standard schedule of rates, issued or published by an employer, will usually list standard items and rates, and the tenderer is asked to submit percentage additions or deductions to reflect current pricing levels. Typical examples are those produced by the PSA and the National Schedule of Rates. Since the tender is a single figure, contractor selection is simple. Schedules of this sort enable orders to be placed before the project details are complete. An *ad hoc* schedule of rates is a pricing document prepared for a particular job. Only those items needed for the project will be incorporated. This type of schedule is difficult to use because, in the absence of quantities, tenders are difficult to compare and the value of the project is not known at the start. An *ad hoc* schedule should contain approximate quantities to help overcome these problems. With all schedules of rates used at tender stage, the estimator is unable to foresee the full extent of the work. Contractors have been asked to quote for drainage trenches, for example, without knowing the ground conditions. Should the contractor assume that the ground conditions were good, and free of obstructions and other services?

Cost-reimbursement contracts

The basis of this method is for the contractor to be repaid with the prime cost as defined in the contract, and a management fee to cover overheads and profit. The fee can be based on a percentage of cost (cost plus percentage contract) or a lump sum based on the estimated project cost (cost plus fixed-fee contract). The advantages of this method are: the project can start quickly, the contractor can contribute to the design, competition can be introduced through the size of the fee, and the contractor is unlikely to cut corners. The disadvantages may be: the contractor has little incentive to save on time and resources (in some management contracts if the construction costs rise the fee to the management contractor also rises), the client is unable to predict the total cost accurately, and it can be tedious to calculate costs during the construction stage. It should be remembered that most of the work is carried out by package contractors who tender for work on a traditional bill of quantities.

Formal tender documents

Formal invitation

The Code of Procedure for Single Stage Selective Tendering gives an example letter. The letter is not long because essential information is normally set out in the tender documents. The letter is needed to tell the contractor which drawings have been sent, arrangements for site visits, date for return of tender and how the tender should be submitted. If the tenderer wishes to decline an offer he should have done so at pre-selection stage. The client should issue the tender documents on an agreed date in order to enable the contractor to plan his estimating workload. A typical invitation to tender letter is given in Fig. 4.1.

John Price & Partners
Chartered Quantity Surveyors
32 Westgate Road
Northbridge NB33 6XD

30th May 1994

C.B.Construction Ltd
8 Brecon Road
Northbridge NB21 8DR

Dear Sirs,

<u>INVITATION TO TENDER</u>
<u>NEW OFFICES FOR FAST TRANSPORT PLC</u>

Following your request to tender for the Fast Transport
contract, we enclose the following documents:

1. two copies of the bill of quantities
2. the general arrangement drawings
3. two copies of the form of tender
4. envelope for the return of the tender

The completed form of tender is to be sealed in the
envelope provided, and sent to the architect's office to
arrive not later than <u>12 noon on Friday 15th July 1994</u>.

The complete set of contract drawings and site
investigation report may be inspected during normal
working hours at the offices of the architect, the
Swallow Partnership, at 102 Cantilever Drive, Stansford.
Arrangements to visit the site should be made with the
project architect, Mrs. K. Edwards tel: 0123 344334.

Please acknowledge receipt of this letter and tender
documents.

Yours faithfully,

Fig 4.1 *Typical formal invitation letter*

Bill of quantities

If a priced bill of quantities is required with the tender then two copies should be sent to each contractor. As much information as possible should be included in the bill to reduce the need for many drawings to accompany enquiries to sub-contractors. If domestic sub-contractors are named in the bill then the consultants can send copies of the drawings and specification direct to the specialists to assist the contractors not least in reducing the reproduction and postage costs.

Drawings

The bill of quantities will list the drawings which were used in preparing the documents. With standard methods of measurement aimed at producing shorter bills of quantities, there is a greater reliance on drawings by the tenderers. Tendering costs could be cut if copy negatives or reduction prints can be produced.

Form of tender

The form of tender is a pre-printed formal offer, usually in letter form, which ensures that all tenders are received on the same basis and should be simple to compare. The tenderer fills in his name and address and a sum of money, for a lump-sum offer. It may be sent with a collusive tendering certificate and appendices which are used for declarations about 'fair wages' or 'basic lists of materials'. A typical tender form is shown in Fig. 4.2, and Fig. 4.3 is an example of an alternative tender which might be produced in addition to a compliant bid.

Return envelope

Each contractor should be provided with a pre-addressed envelope clearly marked 'Tender for . . .'. They are to be marked so that they will be easily recognized and not opened too early or by the wrong person. Some clients insist that the contractor's name must not appear on the envelope.

FORM OF TENDER

To: Fast Transport Ltd, Stansford

Tender for: Proposed Office Building, Stansfield

Dear Sirs,

Having examined the conditions, drawings and bills of quantities, we offer to carry out and complete the works described, for the FIRM price of:

£_____ (in words)_____

_____

and complete within 34 weeks from the date of possession.

This tender will remain open for acceptance for three months from the date of return of tender.

We agree that should any obvious pricing or arithmetic error be discovered before acceptance of this offer in the priced bills of quantities then these errors will be corrected using Alternative 1 in Section 6 of the Code of Procedure for Single Stage Selective Tendering.

We understand that we are tendering at our own expense and that neither the lowest nor any tender need be accepted.

Signature _____ Date: _____

Company: _____

Address: _____

_____

Fig 4.2 *Typical form of tender*

C.B.Construction Ltd
8 Brecon Road
Northbridge
NB21 8DR

15th July 1994

Fast Transport Ltd
Stanton Lane
Stansford

Dear Sirs,

New Offices, Stansford
Alternative Tender

Following discussions with the architect and engineer
during the tender period, we have examined an alternative
design which would lead to a significant saving of time
and money, as follows:

1. By a small increase of plan dimensions (to the lines
    shown on our layout drawing F/1 attached) including
    some accommodation in the roof space, there would be
    no need for the basement construction.

2. You will see on our preliminary programme (our
    drawing number F/2) the contract duration can be
    reduced by 4 weeks to 30 weeks, with completion by
    22nd December.

3. Our alternative proposals would offer a financial
    saving amounting to £29,552 and a tender sum of
    £811,100.47.

We hope that this will help you in your appraisal of the
scheme and would be pleased to provide more information
and discuss the work with you soon.

Yours faithfully,

J.Lewis
Regional Manager

Fig 4.3 *Example of an alternative tender*

Estimating methods

'Brian does his estimates on the back of a cigarette packet'

Introduction

During the first half of the twentieth century six methods of estimating were used (Fig. 5.1). The methods are much the same today. The main difference is the current popularity of elemental cost models which are used by quantity surveyors and contractors alike in advising clients on their likely building costs and helping designers to work within a budget.

The approximate methods of estimating depend on reliable historical cost data whereas the analytical approach to estimating is based on current prices for

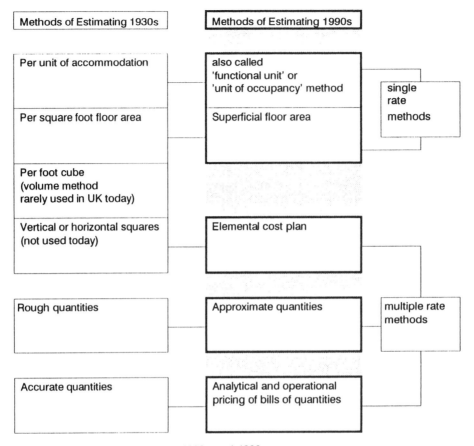

Fig 5.1 *Estimating methods in 1930s and 1990s*

resources. A contractor may use most of the methods in developing a single design and build project. A client, for example, could be given a cost range during a discussion using the unit method and an elemental cost plan would be produced when the client's outline brief is received. Approximate (or builder's) quantities are used to produce a formal tender and when a contractor has received an order, a full bill of quantities may be written for financial control during construction. The use of cost planning methods enables the contractor or PQS to provide the client with informed advice to ensure value for money. The first step is to calculate a budget at the inception of a project and the second a cost plan using preliminary drawings. The contractor is in the unique position of having detailed knowledge

54

of current prices for all the resources used in construction. The PQS has the benefit of rates submitted in priced bills of quantities from a broad selection of contractors, although he must be aware that rates do not necessarily reflect the actual cost of individual items of work.

The final cost of construction may be different from the forecast, for many reasons, namely:

1. The type of building; schools may be easier to predict than a bridge, the extent of repairs in maintenance contracts can be difficult to foresee;
2. The effect of competition in the market;
3. The amount and quality of historical data available;
4. The amount of design information available;
5. The performance of the design team;
6. The nature of the workplace in terms of weather, ground conditions, resource prices and other uncertainties;
7. Changes introduced by the client;
8. The estimator's skill and method used.

The contractor's estimator has the dual roles of forecasting the cost of construction and advising how competing organizations will bid for the same job. Although the commercial tender is the responsibility of management the estimator must tell his managers how market trends will affect the prices, particularly where sub-contracting has a strong influence on the tenders.

Single-rate approximate estimating

Unit of accommodation method

This method is commonly used by national bodies such as the education and health services at the inception stage of construction. If a client has an amount of money to spend (a target cost) then it would be possible to consider the likely number of functional units which can be provided. From experience, it might be found that the cost of providing a bed space in a district general hospital is £40 000. Using this figure an expenditure of £20 million would provide approximately 500 beds. On the other hand, if the number of units is known, a budget cost (usually expressed as a cost range) can be calculated.

Providing there are recent comparable data available, the unit method is useful where a simple and quick cost range is needed in the early stages. It is difficult, however, to adjust the costs for specific projects, in different locations, with varying ground conditions and so on.

Floor area method

The main reason for the popularity of the floor area method is its simplicity. There are few rules to remember and the cost per square metre is well understood by property developers. A proposed building is measured at each floor level using the internal dimensions but no deductions are made for internal walls, stairs or lift zones. Previous similar building costs are used by dividing the construction cost by the internal floor area. Adjustments can be made for location and inflation; but specification adjustments are much more difficult to estimate. Subjective judgements are made for size, shape, number of storeys, services, ground conditions and standard of finishes. A separate assessment should be made for external works, incoming services and drainage which can be significantly different for similar buildings.

There are many buildings where the unit method is impracticable, such as warehouse projects or open-plan offices. In these cases the superficial floor area method is found to be reliable with an accuracy of 10% to 15%. This method also works well with certain external works contracts such as concrete paving or macadam surfacing.

Building volume method

There are several methods which use the volume of a building as the cost yardstick but they are not widely used today. In some European countries, architects and engineers are familiar with building costs expressed as cubic metre prices. In Germany, there are publications which list typical building costs in terms of their volume and the procedure for calculating volumes is given in a DIN standard.

Multiple-rate approximate estimating

Elemental cost plans

A cost plan is prepared from the designer's preliminary drawings. This is a list of the elements of a building such as substructure, frame and upper floors each with its share of the total budget cost (see Fig. 5.2).

The forecast cost of each element can be calculated in two ways:
1. By measuring the approximate size of each element and applying a unit rate;
2. By calculating the proportion of total cost for each element on a similar building and using this ratio to divide the budget for the proposed building into its elemental breakdown.

```
-----------------------------------------------------------------
CB Construction Limited                            Northbridge
-----------------------------------------------------------------

Proposed Workshop for Fast Transport Limited     2310 m2

-----------------------------------------------------------------
Element                                            Element
                                                   cost
-----------------------------------------------------------------

Substructure                                       95,893

Superstructure

      Frame                                        89,980
      Roof coverings                               47,535
      Roof drainage                                 5,124
      External walls                               44,182
      Windows                                      16,887
      External doors                                6,788
      Internal walls                                3,834
      Internal doors                                3,377

Internal finishes

      Wall finishes                                 7,125
      Floor finishes                                4,521
      Ceiling finishes                              1,846

Fittings and furniture                             2,355

Services

      Sanitary appliances                           3,517
      Internal drainage                              inc
      Hot and cold water                             inc
      Heating                                       2,990
      Electrical installation                      10,433
      BWIC                                            950

External works

      Site works                                   51,233
      Drainage                                     11,482
      External services                             2,000

Preliminaries                                      62,247

Contingencies                                      22,000

-----------------------------------------------------------------
Budget total                                    £ 496,299
-----------------------------------------------------------------
```

Fig 5.2 *Elemental cost plan for portal-framed building*

The second method is better shown by example. If a contractor has built some portal-framed factories he will know the costs of each element and can express this information as costs for each unit of floor area. Figure 5.3 illustrates a typical analysis for a factory building.

```
--------------------------------------------------------------
CB Construction Limited                              Northbridge
--------------------------------------------------------------

Factory for Hitech Cables Limited                      3120 m2

--------------------------------------------------------------
Element                        Element cost           Cost/m2
--------------------------------------------------------------

Substructure                        109,822             35.20

Superstructure

    Frame                           120,905             38.75
    Roof coverings                   69,735             22.35
    Roof drainage                     6,550              2.10
    External walls                   55,692             17.85
    Windows                          15,289              4.90
    External doors                    9,826              3.15
    Internal walls                    5,143              1.65
    Internal doors                    8,895              2.85

Internal finishes

    Wall finishes                    10,296              3.30
    Floor finishes                    5,614              1.80
    Ceiling finishes                  3,430              1.10

Fittings and furniture                4,365              1.40

Services

    Sanitary appliances               4,520              1.45
    Internal drainage               inc
    Hot and cold water              inc
    Heating                          13,884              4.45
    Electrical installation          21,060              6.75
    BWIC                              2,189               .70

External works

    Site works                       72,850             23.35
    Drainage                         18,400              5.90
    External services                 2,960               .95

Preliminaries                        84,240             27.00

Contingencies                        32,000             10.25

--------------------------------------------------------------
Totals:                        £ 677,665            £ 217.20
--------------------------------------------------------------
```

Fig 5.3 *Elemental cost plan for building under construction*

A cost plan for another similar factory can be generated by multiplying each rate by the new floor area. Figure 5.4 shows the second factory which the contractor will further adjust for inflation, and significant specification changes. Typical examples would be the number of sanitary appliances, internal doors, roller shutter doors and ground improvements. In this example the contractor was confident about this approach because he found the floor area and wall-to-floor ratio to be similar to the earlier factory. If a budget is wanted for another factory with a much smaller floor area, say 1200 m², then a different approach would be needed. Since the wall/floor ratio will be greater, the estimator should look at the area of walling and apply a rate/m². The preliminaries cannot be assessed using the floor area either. An allowance for preliminaries should be calculated using the cost per week for a similar factory and multiplying by the duration. In this way, a combination of historical data (the cost of elements per m² of floor area) and calculated costs for certain elements is used.

Contractors and PQSs are becoming more adept at using this method and have adapted the basic principles for computer systems. A spreadsheet template can store the information shown in Fig. 5.4 and the effect of changes can be seen immediately they are made. In fact computers are now used to produce sophisticated budgets for clients at the early stages of design.

Approximate quantities

There are many ways in which approximate quantities are used, depending on who uses them and for what purpose. A PQS may want an alternative estimating technique to check cost forecasts before tenders are returned. Measurements will be concentrated into as few items as possible for grouped work components. A simple example is a cavity wall measured and priced with both skins included in the unit rate, with forming the cavity, wall ties, plastering and pointing. Rates for composite items can be found in price books, calculated from rates in priced bills of quantities or from first principles.

A contractor needs to produce bills of approximate quantities when tendering for work based on drawings and specifications. He will seldom allow all the ancillary and subsidiary work items found in the standard method of measurement; but must be careful to tell sub-contractors the assumptions made. There is a strong case for attaching a preamble on the rules of measurement used, so that any misunderstandings and disputes will be reduced (Fig. 5.5).

The accuracy of this method is related to how far the design has developed. At least the quantities are based on the planned construction and not a previous job and allowances are made for plan shape, height of building, type of ground, quality of finishes, etc. For these reasons it is widely used and being developed

59

```
-------------------------------------------------------------------
CB Construction Limited                      Northbridge
-------------------------------------------------------------------
                        Hitech                Pluto
                                              Blinds
                        3120m2                2860m2
           ----------------------------------------------------
Element                         Cost/m2    New budget
           ----------------------------------------------------

Substructure            109,822    35.20    100,672

Superstructure

     Frame              120,905    38.75    110,825
     Roof coverings      69,735    22.35     63,921
     Roof drainage        6,550     2.10      6,006
     External walls      55,692    17.85     51,051
     Windows             15,289     4.90     14,014
     External doors       9,826     3.15      9,009
     Internal walls       5,143     1.65      4,719
     Internal doors       8,895     2.85      8,151

Internal finishes

     Wall finishes       10,296     3.30      9,438
     Floor finishes       5,614     1.80      5,148
     Ceiling finishes     3,430     1.10      3,146

Fittings and furniture   4,365     1.40      4,004

Services

     Sanitary appliances  4,520     1.45      4,147
     Internal drainage    inc
     Hot and cold water   inc
     Heating             13,884     4.45     12,727
     Electrical installation 21,060 6.75     19,305
     BWIC                 2,189      .70      2,002

External works

     Site works          72,850    23.35     66,781
     Drainage            18,400     5.90     16,874
     External services    2,960      .95      2,717

Preliminaries           84,240    27.00     77,220

Contingencies           32,000    10.25     29,315

-------------------------------------------------------------------
Totals:               £ 677,665 £ 217.20  £ 621,192
-------------------------------------------------------------------
```

Fig 5.4 *Elemental cost plan for similar factory building*

```
-------------------------------------------------------------
Factory for Hitech Cables Limited              February 1994
-------------------------------------------------------------
        Description                          quant    unit
-------------------------------------------------------------

a.      Excavate to reduce level               332     m3

b.      Excavate for foundations ne 1.0m deep  248     m3

c.      Excavate machine pits ne 4.0m deep     112     m3

d.      Load and remove surplus from site      445     m3

e.      Backfilling with selected exc material 247     m3

f.      DOT type 1 under slab 400mm thick      1330    m3


        -----------------------------------------------------
        Total to summary
        -----------------------------------------------------




-------------------------------------------------------------
Factory for Hitech Cables Limited               January 1994
-------------------------------------------------------------

PREAMBLE TO BILL OF QUANTITIES

        -----------------------------------------------------

EXCAVATION WORKS
----------------

The following information is provided to outline the location and
layout of the excavation works:

a.      Drawing 3409/1          Site Plan

b.      Drawing 3409/2          Details of foundations

c.      Drawing 3409/4          Details of ground slab and beam

d.      Drawing 3409/7          Details of machine pits

e.      Ground investigation report


The excavation work has been measured under the following headings;
the item coverage is described in the right-hand column:
```

Fig 5.5 *Example of 'builder's quants' and associated preamble*

<u>ITEM COVERAGE</u>

Excavation rates to include:

1. Excavation in any type of subsoil to the depths shown on the drawings

2. Dealing with surface water affecting the excavations

3. Dealing with ground water entering the excavations

4. Excavating over or around existing services

5. Any extra width of working space needed for sub-structure work

Excavate to red level

1 The quantity of excavation includes an allowance of 500mm for working space from the outside face of external walls

2 The depth of excavation is not stated

3 Any necessary earthwork support is deemed to be included in the rates

4 Levelling and compacting the ground is deemed to be included in the rates

Fig 5.5 *continued*

with computer systems using database and spreadsheet software to produce standard bills for repetitive building types. The danger is that the cost calculated using approximate quantities can appear to be as accurate as a full bill of quantities based on working drawings. It is more likely to be an underestimate of the cost of construction unless a generous contingency is added for small components, fittings, fixings and design development.

In common with all approximate estimating techniques there are some difficulties which need to be recognized when advising clients. Some of the difficulties to be faced are:

1. The reliability of historical data must always be questioned.
2. Preliminaries are usually unique to a particular job and should be calculated whenever there is deviation from an identical scheme.
3. Incoming services are seldom the same on different sites and can only be assessed after detailed consultation with service providers.

4. Contract conditions can vary markedly between projects; the requirements for bonds, insurances and liquidated damages can be particularly onerous.
5. The contingency sum for design development must be estimated for each job.

Analytical estimating

Analytical estimating is a method for determining unit rates by examining individual resources and the amounts needed for each unit of work. This method for pricing bills of quantities is described in the CIOB Code of Estimating Practice, in three stages:

1. Establish all-in rates for the individual resources in terms of a rate per hour for labour, a rate per hour for items of plant and the cost per unit of materials delivered and unloaded at the site.
2. Select methods and outputs to calculate net unit rates to set against items in the bill of quantities.
3. Calculate project overheads, summarize resources and prepare report for management.

The ability to analyse unit rates is an important skill for all those engaged in construction. Quantity surveyors and architects may need to value variations using clause 13.5.1.3 of the Standard Form of Building Contract. This states 'where the work is not of similar character to work set out in the contract bills the work shall be valued at fair rates and prices'. This presumably means a properly built-up unit rate. Contractors rely on the pricing carried out by their sub-contractors for an increasing share of the work. A contractor's estimator should be able to build up rates for his direct work and be able to check the rates offered by sub-contractors.

Analytical pricing of bills of quantities is more than just applying resources to items of work to produce a unit rate. The constituents of a rate are inserted in the bill; and totalled for each page, each section, and carried to the summary, so that the contractor has a complete picture of the resource costs at the adjudication meeting. Figure 5.6 shows a typical printout from a contractor's bill where the rates and totals are shown between the item descriptions. Figure 5.7 is an example of a contractor's bill of quantities priced analytically using a spreadsheet package.

The benefits of analytical pricing of bills of quantities are:

1. The total cost of labour is needed to calculate the cost of insurances, transport of operatives, small tools and equipment, and workforce levels.

```
-------------------------------------------------------------------------
Factory for Hitech Cables Limited                   February 1994

-------------------------------------------------------------------------
       Description                           quant unit      rate   total
-------------------------------------------------------------------------
       LAB        PLT         MAT        S/C
-------------------------------------------------------------------------

a.     Excavate to reduce level              332   m3    3.50    1162.00

       1.00        2.50
       332         830


b.     Excavate for foundations ne 1.0m deep 248   m3    4.60    1140.80

       1.50        3.10
       372         769


c.     Excavate machine pits ne 4.0m deep    112   m3    5.80     649.60

       2.00        3.80
       224         426


d.     Load and remove surplus from site     445   m3    6.90    3070.50

                   5.00        1.90
                   2225        845


e.     Backfilling with selected exc material 247  m3    3.80     938.60

       1.00        2.80
       247         692


f.     DOT type 1 under slab 400mm thick     1330  m3    15.90  21147.00

       1.00        2.80        12.10
       1330        3724        16093


-------------------------------------------------------------------------
       Total to summary                                       28108.50

       2505        8666        16938                          28109
-------------------------------------------------------------------------
```

Fig 5.6 *Contractor's bill of quantities priced analytically*

64

Weighbridge foundation for Hitech Cables Limited   13.2.94

Analysis of rates

Surface weighbridge (15m)	quant	unit	rate	total	lab	plt	mat	s/c	LAB	PLT	MAT	S/C
A Excavate to reduce lev ne 1.0m dp	23	m3	5.40	124.20	2.40	3.00			55	69		
B Excavate for thickening & downstand	9	m3	15.00	135.00	12.00	3.00			108	27		
C Load and remove to tip on site	17	m3	4.80	81.60		4.80				82		
D Backfill with selected material	15	m3	3.40	51.00	1.40	2.00			21	30		
E Level and compact	79	m2	0.45	35.55	0.35	0.10			28	8		
F Earthwork support	6	m2	1.95	11.70	0.65	1.30			4	8		
G Hardcore (Free Issue)	18	m3	3.40	61.20	1.40	2.00			25	36		
H Blind with dust (Free Issue)	62	m2	0.40	24.80	0.20	0.20			12	12		
J Soil stabilization mat	79	m2	0.80	63.20	0.25		0.55		20		43	
K Concrete grade 40N in foundation	17	m3	58.00	986.00	14.00		44.00		238		748	
L Concrete grade 40N in ramps	15	m3	60.00	900.00	16.00		44.00		240		660	
M Concrete grade 40N in upstands	6	m3	78.00	468.00	30.00		48.00		180		288	
N Concrete grade 40N in plinths	1	m3	73.00	73.00	25.00		48.00		25		48	
P Rebar 12mm (upstand & downstand)	0.6	t	645.00	387.00	215		430		129		258	
Q Rebar 16mm	0.12	t	600.00	72.00	200		400		24		48	
R Fabric A393	280	m2	4.10	1148.00	0.85		3.25		238		910	
S Dowel bars: 25mm	30	nr	2.00	60.00	1.00		1.00		30		30	
T Form plinths 900x900x250mm high	4	nr	19.50	78.00	13.00		6.50		52		26	
U Sawn formwork to sides of founds	47	m2	21.00	987.00	14.00		7.00		658		329	
V Sawn formwork to sides of upstands	17	m2	23.00	391.00	15.00		8.00		255		136	
W Cast in service duct	1	nr	10.00	10.00	6.00		4.00		6		4	
X												
Y Grouting baseplates on return visit	1	item	155.00	155.00	125		30		125		30	
Z Steel bumper stops	284	kg	1.90	539.60	0.40		1.50		114		426	
TOTAL				6842.85								
check				6843				6843	2587	272	3984	

Fig 5.7 *Contractor's bill for weighbridge foundation*

65

2.   The breakdown of resource costs is needed to calculate the allowance for firm price tenders.
3.   Labour and plant totals for elements of the work are used to calculate activity durations for the tender programme.
4.   A breakdown of prices is needed in each trade to make comparisons between direct work and labour only sub-contracts.
5.   The costs of resources are needed to calculate the cost commitment cash flow forecast.
6.   Adjustments can be made to any part of the estimate right up to the submission date.
7.   The resource breakdowns will be used on site for post-tender cost control, bonus systems, monitoring and forward costing.

Most contractors know the benefits but have difficulty in practice. The two main problems are that all the rates must be priced analytically for the system to work, and many extra calculations are needed to extend the rates to totals. A computer estimating system is designed to overcome these difficulties and will produce the resource summaries automatically.

Unit rate pricing of a bill of quantities is carried out to certain conventions; those which are expected by the client's representative and those which the contractor has developed. The notes at the beginning of a bill of quantities usually include instructions such as:

1.   All rates shall be inclusive of labour, materials, transport, plant, tools, equipment, establishment and overhead charges, and all associated costs, margins and profit.
2.   All items shall be priced; the value of any items unpriced shall be deemed to be included elsewhere in the bill of quantities.

The contractor, on the other hand, is likely to produce rates which exclude:

1.   General overheads and establishment charges;
2.   Profit, which can only be calculated after the net estimate is complete;
3.   Restrictions which apply to more than one item such as difficult access, difficult handling and protection;
4.   Plant which is common to several activities such as compressors, hoists, mixers, dumpers and cranes.

Contractors may include a nominal mark-up or 'spread' to the rates, which can be supplemented by sums in the preliminaries part of the bill when the true overheads and profit are known after the adjudication meeting. A computer-aided estimating system would allow some of the overheads and profit to be spread over various parts of the bill of quantities. For example, a contractor might want to add 20% to the earthworks and 10% to the painting work. This could improve the cash-flow position of the project but would put the contractor at risk if the extent of earthworks reduced.

There are many PQSs and civil engineers who would want to introduce analytical bills. This would be a bill format with extra columns for labour, plant, materials, sub-contractors, and overheads/profit, which would be submitted by the lowest tenderer before entering into a contract. The client's consultants argue that although contractors may resist this duty to reveal confidential information, the idea has the following advantages:

1. There would be a clearer basis on which to value variations.
2. The settlement of final accounts could be based on an examination of which elements had changed, and the effect on the programme may be clearer.
3. The design team could see where they had chosen designs which were labour-intensive.
4. If the analysis was extended into valuations, the contractor could use the data for his own cost-monitoring systems without doubling his effort.

There may be contractors who will object to giving a full breakdown of their rates and there will be others who will insert all their rates in the sub-contract column. Change will come, however, from two directions: the exchange of tender documents using computer files and trust brought about by partnerships between clients and contractors.

Operational estimating

Operational estimating is a form of analytical estimating where all the resources needed for part of the construction are considered together. For example, an estimator pricing manholes using the Civil Engineering Standard Method of Measurement needs to gauge the time taken to build a complete manhole, whereas a building estimator is expected to price all the individual items for excavation, concrete work, brickwork, etc., measured under the rules of the appropriate work sections.

The following examples show some of the many other situations where work is priced as whole packages:

1. Excavation including trimming, consolidation and disposal;
2. Placing concrete in floor slabs including fabric reinforcement, membranes, isolation joints and trowelling;

3. Formwork to complex structures including a unique design, hired-in forms and falsework;
4. Drain runs including excavation, earthwork support, bedding, pipework and backfill;
5. Repairs which often involve more than one trade or a multi-skilled operative;
6. Roof trusses including the use of a crane, a suitable gang of operatives and temporary works.

It must be said that building estimators have become skilled at applying production outputs to units of work and then occasionally employing operational estimating techniques to check the results. Civil engineers, on the other hand, usually examine methods and durations before pricing the work. This is because different construction methods for civil engineering can have a significant effect on costs. There is also a greater reliance on the specification, the drawings and preambles which give the item coverage.

The term 'operational estimating' is often applied to methods which rely on a forecast of anticipated durations of activities and a resource-levelling exercise. The estimator must start with an appraisal of the details on the drawings, the extent of the work described in the specification and bill, and a study of the site conditions. Next, the sequence of work will be found by considering the restraints brought about by site layout, client's requirements, the design, time of year, and temporary works. The critical operation at each stage of the construction can then be plotted and the rest of the activities sketched in. Labour and plant schedules can be drawn up for direct work and specialist sub-contractors will be asked for advice about their work. It may be necessary to change the programme if there are any unwanted peaks and troughs in the resources needed on site. The estimator will then have a list of resources for each operation from which to calculate costs. This approach will often produce a cost based on a particular method for carrying out the work. If this has brought about a saving in costs the estimator will prepare a method statement so that site staff can understand the assumptions made in preparing the estimate.

When a building estimator uses operational estimating with a traditional bill of quantities he has great difficulty dividing the cost of a piece of work among all the related bill items. Where, for example, should an estimator put the rate for casting a concrete floor which includes a DPM, fabric reinforcement, power floating and sealer? The PQS often insists on rates being inserted against items which have a value so that there is better financial control during construction. Clearly this is not a problem with a contractor's bill of quantities produced for design and build or plan and specification projects. Another solution to the problem for building estimators would be to rough price the bill early in the tender period and adjust the balancing sums of money when operational methods highlight greater or lower costs. This is commonly done during the adjudication stage, and the rough pricing technique is popular with those using computer systems.

The advantages of operational estimating are:

1. Activities are examined to select those which are practicable.
2. Outputs are based on a programme, which includes holiday breaks, time of year, idle time, facilities available on site, etc., giving a more realistic guide to the time needed for labour and plant.
3. Alterations and repair work are usually measured as global items which can be overpriced if all the possible trades are examined separately.
4. In a competitive market, the estimator may only look at the labour and plant needed for the core item of work; such as the brickwork in a manhole assuming the bricklayer can fix the cover while finishing the brickwork and the excavator can dig the pit when it digs the pipe trench.

Figure 5.7 is a contractor's bill of quantities for a weighbridge foundation priced analytically. The estimator used an all-in rate of £7.00/hour for all his labour and used his usual labour outputs from his tables of constants.

The site manager has kept records from previous similar jobs which show that this type of weighbridge foundation usually takes two weeks to construct with four men, and a return visit is needed for two men to grout in the equipment. A backacter and roller costing £18.00/hour is needed for three days. This gives the following net cost for labour and plant:

Labour	4 nr × 2 weeks ×	45 hours × £ 7.00 =	2520.00
	2 nr × 2 days ×	9 hours × £ 7.00 =	252.00
		Total	£2772.00

Plant	1 nr × 3 days ×	9 hours × £18.00 =	£486.00

It can be seen from the comparison that when the project is assessed as a whole, the net cost of construction is marginally more than the total from the unit rate analysis. The estimator may have used his normal constants for labour and plant without checking whether there is a continuous flow of work for labour and plant resources. Perhaps the site manager should next look at materials wastage which he has experienced, in particular blinding concrete and fabric reinforcement which could be significantly higher for such a small contract.

Contractor selection and decision to tender

Introduction

How does a construction organization maintain its turnover? Some enquiries arrive 'out of the blue' arising from hearsay, the *Yellow Pages*, or advertising. Others are sent on the strength of earlier successful contracts or following a direct sales approach. New markets can be entered by replying to invitations for open tenders; some opportunities can be created by speculation. Currently the greater part of work carried out in the construction industry is secured through a process of tendering which is intended to be an unbiased means of selecting a contractor to carry out work. The criteria for selection are determined by the client through an evaluation of his needs. The aims of selection are to find a contractor who can supply a product for the lowest possible price, and can demonstrate the following:

1. A reputation for good quality workmanship and efficient organization.
2. The ability to complete on time.
3. A strong financial standing with a good business record.
4. The expertise suited to the size and type of project.

The construction industry is concerned with more than providing off-the-shelf products; most projects involve unique designs with purpose-written specifications to be finished in a time which is often difficult to predict. Construction clients must balance the importance of cost, quality and time because it is rare for all three to be satisfied. A client can reduce the financial risks by fully designing the project before selecting contractors.

It is not only clients who need to establish the financial standing of the other party. The contractor must be satisfied that the client has the ability to pay, and on time. In the past contractors have not been so careful about selecting their clients. This has now changed with the introduction of bonds and guarantees which are used by both parties to contracts. The word 'trust' is unfortunately fast disappearing from conventional agreements, and lawyers are often the main beneficiaries. Nevertheless, contractors are always keen to be invited to tender

71

where competition is restricted to a selective list of tenderers, and are getting involved in more innovative relationships such as 'turnkey' and 'project partnerships'.

Competition and negotiation

Contractors may be selected by competition or negotiation and sometimes by a combination of both. Open competition is an arrangement where an advertisement in local newspapers or trade journals invites contractors to apply for tender documents. A deposit is usually required to ensure that only serious offers are made; presumably it is needed to cover the cost of copying the documents. Local authorities have been advised against open tendering because it often leads to excessive tender lists where the cost of abortive tendering is considerable. There are instances of selection criteria being applied after the tender has been submitted, so a bid could be rejected if a contractor does not belong to an approved trade association, for example, after he has submitted his tender. They argue that this method allows new contractors to join the market and increases the chance of gaining a low price. The larger contractors avoid this method because they can see no reason to compete against anyone who asks to be included on the tender list and later be subjected to the further hurdle of contract compliance clauses.

Selective tendering consists of drawing up a list of chosen firms and asking them to tender. It is by far the most common arrangement because it allows price to be the deciding criterion; all other selection factors will have been dealt with at the pre-qualification stage. There are three ways in which selective tendering lists are drawn up:

1. An advertisement may produce several interested contractors and suitable firms are selected to tender.
2. The consultants may contact those they would wish to put on an *ad-hoc* list.
3. Many local authorities and national bodies keep approved lists of contractors in certain categories, such as work type and cost range.

Contractors who ask to be included on select lists of tenderers are usually asked to provide information about their financial and technical performance, particularly about the type of work under consideration. The National Joint Consultative Committee for Building (NJCC) has written the 'Standard form of tendering questionnaire – private edition' so contractors can prepare answers to relevant questions in advance. The questions mainly deal with projects carried out during the previous three years. Once the form has been completed, it can be used for specific projects or for those compiling lists of selected contractors.

The building industry follows the 'Code of Procedure for Single Stage Selective Tendering' also published by the NJCC. This procedure follows a number of well-defined stages for pre-selection and tender stage actions. Its success relies on complete designs before tenders are invited and the use of standard forms of contract but can be used with other procurement systems. The following points illustrate the coverage of the code:

1. Preliminary enquiry – contractors are given the opportunity to decide whether they wish to tender by receiving a preliminary enquiry letter, 4 to 6 weeks before the despatch of tender documents.
2. Number of tenderers – the recommended number of tenderers is a maximum of six (three or four for design and build) and further names could be held in reserve.
3. Tender documents – the aim of the documents is that all tenders will be received on the same basis so that competition is limited to price only.
4. Time for tendering – normally at least 4 working weeks should be allowed, and more time may be needed depending on the size and complexity of the project.
5. Qualified tenders – tenderers should not try to vary the basis of their tenders using qualifications. Queries or unacceptable contract conditions should be raised at least 10 days before tenders are due. The consultants can then tell all the tenderers of their decisions and if necessary extend the time for tendering. A contractor should be asked to withdraw significant qualifications or else face rejection. This is necessary to ensure tenders are received on a like-for-like basis.
6. Withdrawal of tenders – a tender may be accepted as long as it remains open; a definite period is usually stated in the tender documents. The tenderer may withdraw his offer before its acceptance, under English law.
7. Assessing tenders – the tenders should be opened as soon as possible after they are received. Priced bills may be submitted in a separate envelope by all the contractors, or, more likely, only the bills of the lowest tenderer will be called for and submitted within four working days. Once the contract has been let, every contractor should be issued with a list of tender prices. Alternatively, tender prices should be given in ascending order and the names listed in alphabetical order.
8. Examination and adjustment of priced bills – the PQS will treat the information in the tender documents as confidential and report errors in computation to the architect and client. There are two methods for dealing with errors. Alternative 1 gives the tenderer the opportunity to confirm his offer or withdraw it. Alternative 2 allows the contractor to confirm his offer or amend it to correct genuine errors. If the contractor amends his offer with a revised tender which is no longer the lowest, the tender of the lowest will be considered.

73

9. Negotiated reduction of tender – the code of procedure recognizes the need to look for savings in the cost of a project where the tender exceeds the employer's budget. This can be achieved by negotiation with the lowest tenderer, or the next lowest if negotiations fail.

Two-stage selective tendering may be adopted as an alternative to single-stage selection when a contractor's assistance is needed during the design stage. The first stage will produce a competitive tender based on approximate bills of quantities using preliminary design information. The contractor selected at the first stage helps with design, programming, and cost comparisons, and submits a final tender for the works, without competition, based on the original pricing levels.

The National Joint Consultative Committee for Building has published codes of procedure for two-stage selective tendering and selective tendering for design and build. The principles are the same as those described for single-stage tendering. For design and build schemes the client must ascertain the design and build experience of each contractor and limit the number of tenderers to three, or four at most, because there are large costs involved in preparing designs and cost proposals. Contractors must be told the basis for assessment where the price is not the sole basis for the award. The code suggests that the relative importance of cost, quality and time for construction should be included in the Employer's Requirements. An employer could, for example, state the target cost and time-scale in his tender documents so the principal criteria for selection will be the quality and appearance of the building.

When a contract is negotiated, a contractor is often selected on the basis of past performance, recommendation, familiarity with the work, or from previous experience with the client or his advisers. In certain circumstances only one contractor may be able to provide the service required as in the case of system building. It is more difficult for those in the public sector to negotiate because EC directives insist that projects over a specified value must be subject to competition. Negotiation allows early contractor selection where the extent of work is not fully known and time is of the essence, and more time would be wasted in preparing full tender documents.

The process of negotiation starts with an outline design and a pricing document such as a bill of approximate quantities. The contractor will insert rates which will be agreed by negotiation between the PQS and contractor's QS or estimator. Without competition the initial price may be higher than would be gained by other means, but this may not be a serious problem. An employer is often looking for other factors such as confidence, reliability, speed and experience of working with a known contractor.

Serial tenders allow several similar projects to be placed with a particular contractor and thereby provide the incentive of a continuous flow of work. The contractor is normally selected using a priced master bill. Separate contracts for each individual project can then be arranged using the priced master bill as a basis for pricing levels.

Abuse of tendering procedures

The NJCC Codes, and Practice Notes, have encouraged all those involved in tendering to use fair and efficient methods which are the best and most professional techniques in use today. The prime aim is to select the right contractor who will give the client good value for money. Unfortunately, individual interests and lack of time can stand in the way of good practice, and the parties to a contract are often unclear about the true nature of the agreement. Some of the problems faced by the estimator are:

1. *Large tender lists* Open competition has been widely criticized in the construction industry, but it continues to be used, mainly by local authorities. They argue that there cannot be the suspicion of favouritism and the lowest possible price will be secured. They fail to recognize the advice of every committee and working party which has looked at this matter since the early 1940s. The reports of the Simon (1944) and Banwell (1964) committees stress the need to avoid the temptation to rely on price alone; there should be a sensible number of competent firms selected who can comply with the quality and time requirements. Some clients impose performance bonds to make up for the failings in the system, thus adding to the cost of construction and hoping that a poor job can be corrected when a contractor fails to complete satisfactorily. With the high costs of tendering in mind, many reputable contractors will not willingly take part in open tendering, particularly where local authorities have been known to receive tenders from over 30 contractors. In one example an authority issued tender documents to 28 firms interested in tendering for a multi-storey car park using the design and build system. It is difficult to understand the logic of so many architects producing designs with such a small chance of success.

2. *Short tender periods* The time for tendering should be determined by three factors: the size of the project, the complexity of the project and the standard of the documents. In practice the design and tender documentation is often late with clients wanting to make a start on site quickly, thus eroding the time available for the estimate. A 'rough' estimate could be produced quickly but a contingency sum would be needed for unknown risks. Contractors would prefer to examine the project, the site, the documents and agree methods with the contract staff and sub-contractors, prepare a programme and look for

75

tipping facilities. In fact the longer the tender period, the more likely it is that the contractor will find savings which would increase the possibility of winning the contract and may produce a better price for the client. The estimator will try to respond to such short tender periods by telephoning his enquiries to suppliers and sub-contractors, making use of information from previous jobs, manually or with the help of a computer. The depth of analysis will be reduced, there is a greater risk of errors and the price is likely to be greater to reflect such problems.

3. *Tender documentation* The estimator should receive enough drawings to understand the nature and scope of the works. The minimum needed are elevations and floor plans to measure temporary works (such as scaffolding), site plans to consider materials access and distribution and component drawings where non-standard elements are to be priced.

References to brand names and specialist suppliers should include current telephone numbers and addresses. Information must be provided about any restrictions which might affect the contractor's choice of method. The site investigation report (or extracts) should be sent to each contractor. With design and build projects, problems have arisen when all contractors have been expected to carry out their own site investigations – clearly an enormous waste of effort and a further burden on the already considerable costs of tendering.

Perhaps differences between documents might be expected at this stage, and so the bills of quantities are used to specify the amount and quality of the works. Discrepancies between the bill descriptions and specification clauses do cause problems but should reduce with the use of coordinated project information. There will always be people who want to change the agreed conventions. The estimator needs to be alert to traps such as: 'earthwork support shall include all means of holding up the sides of excavations including sheet piling' (normally measurable) or 'hack off external render where necessary and renew' (where necessary could be small isolated sections or the whole wall if the contract administrator so decides). Amendments to the tender documents should be avoided but can be allowed early in the tender period. Once quotations have been received from suppliers and sub-contractors, changes will be difficult to build into the bid.

Estimating without bills of quantities is much more time consuming, not only the time for taking off quantities but enquiries to sub-contractors are delayed and the risk of errors is greater.

4. *Asking for tenders when the work is unlikely to proceed* There is a tradition in the construction industry for estimates to be given without charge to the client. This can be at great cost to unsuccessful contractors. Some have reported that it costs about 0.25% of the tender price to prepare a bid for a traditional lump-sum form of contract: a design and build tender can cost as

much as 2%. Contractors will continue to accept this financial risk providing they are submitting tenders to clients who use selective tendering and eventually award a contract to one of the bidders.

5. *Qualified tenders and alternative bids* The tenderer should submit his bid without adding conditions to his offer. All contractors must consider the terms of their offers, however, and sometimes will not be able to comply fully with the instructions of the client. On the other hand, they should recognize the need for a common basis from which the best bid can be selected. Contractors may produce an improvement to the design or see a method for completing more quickly, and often can calculate an alternative price. Providing an offer is made which complies with the original brief, alternative tenders are considered by employers.

6. *Failure to notify results* A contractor can monitor his tender effectiveness when he receives information about his performance in relation to other tenderers. Tender prices should be published if a contractor is to review his suitability for the type and value of projects.

7. *Late receipt of tender documents* Estimators do their best to deal with requests for tenders sometimes at short notice, but when tender documents arrive later than promised their programme of work will be halted, and other opportunities to tender may have been lost.

Decision to tender

Invitations to tender arrive at a contractor's office in a variety of ways and it is important that they should be channelled to a central source for collating and monitoring. Where the organization has a marketing section then this may be the most suitable location. Alternatively, they can be held within the estimating department. All employees of the firm should be made aware that they have a part to play in capturing the opportunities that arise. Senior management will feed back knowledge of projects gained from conversations with prospective clients and partners in related professions at business and social events. Equally, a job surveyor may well gain knowledge picked up while having a pint with his opposite number from the PQS office. All such snippets of information should be fed to the central source and recorded. A list of expected tenders may become a formal report in bigger organizations so resources can be used effectively.

Formal invitations to tender are normally communicated by either letter or telephone. It is to be hoped that the enquiry follows the format laid down in the Code of Procedure for Single-Stage Selective Tendering (CPSSST) and communicated by letter or facsimile. Compliance with the recommendations given in the Code should be honoured by all parties. The client's professional adviser should provide in good time basic information about the project and ask the contractor if he wishes to be considered for a selective tender list.

The contractor then has the opportunity to decide, knowing there will be a limit on the number with whom he is likely to be in competition.

Regrettably, the telephoned enquiry persists! The person answering the phone needs to ask for all the information he would have if a preliminary enquiry had been sent, as detailed in Appendix A of the CPSSST. He is sometimes asked for a decision immediately, which is usually 'Yes' because he knows that his boss can reverse the decision when the documents come in. It is suggested that a pad of forms be available by the telephone of all those likely to accept a call asking if the firm is willing to tender (see Fig. 6.1). The form contains some basic headings as an *aide-mémoire* to those receiving the request. An abstract from the forms and formal letters of invitation could be kept on a weekly report form.

The decision to tender should be made by the chief estimator or general manager using the following points:

1. Is the work of a type which the contractor has experience, both in winning tenders and completing profitably? Does it conflict with the company's objectives and future workload?
2. Has the contractor the necessary supervisory staff and labour available. He may not wish to recruit untried and unknown personnel in key positions.
3. Will the estimating department have staff available with suitable expertise for the type of work to be priced?
4. Does the location of the proposed site fit the organization's economic area of operation?
5. Are there too many risks in the technical and contractual aspects of the project?
6. Will suitable documents be produced for tender purposes? A busy estimating office may give priority to work which has been measured. Poor documentation might give a clue to the standard of working documents during construction.
7. Has enough time been given to prepare a sensible estimate?

As a client needs to establish the contractor's financial standing, so in turn the contractor will need to be satisfied that the client can pay, and on time. Similarly, as the contractor is investigated for performance on similar work, whether his management structure is satisfactory and his present resources can cope with the added workload, so too the contractor will need to consider experience of working with the architect, engineer, or quantity surveyor.

In Appendix A of the CPSSST the draft letter – Preliminary enquiry for invitation to tender – states: '. . . Your inability to accept will in no way prejudice your opportunities for tendering for further work under my/our direction . . .'.

CB CONSTRUCTION
PRELIMINARY TENDER ENQUIRY

Job title :	Location :
	Value :
Employer :	Architect :
Engineer :	QS :

Brief description :

Form of contract :	Bills :	yes / no
	Fluctuations :	firm / fluct
	Bond :	yes / no
	Damages :	£ per
	Nominations :	

| Programme : | Tender due in : | Start date : |
| | Tender due out : | Duration : |

Action taken :

Comments :

| Signed : | Date : |
| Approved : | Date : |

Fig 6.1 *Preliminary enquiry information form*

79

This is a plea for the contractor to give an honest answer without fear of being penalized in the future. The Code also states that a contractor, having signified initial agreement to tender, should honour that acceptance except in exceptional circumstances. The exceptions are not indicated but it would be reasonable to withdraw if the contractor experienced a sudden increased workload or the documents arrived later than expected. Appendix B of the Code gives the wording for the letter sent to the contractors selected to tender. The model letter starts: 'Following your acceptance of the invitation to tender . . .'. This is a loaded statement, as all the contractor has done is to study the preliminary invitation setting out the basic facts of the proposed project and agreed to be considered for selection. Now he is told he has accepted and here are the documents. Not withstanding the onus placed on the contractor to honour this obligation, now the full documentation is in his possession, he still has every right to confirm or decline to tender.

If the invitation to tender is to be declined, the client's adviser should be told immediately, preferably by phone giving the reasons, and the documents must be returned quickly so that another bid can be invited from a firm on the reserve list. If the decision is to proceed, the estimator should acknowledge the safe receipt of all the tender documents and confirm that a tender will be submitted.

Inspection of tender documents

The arrival of the tender documents within a contractor's office invariably causes a stir; everyone is eager to have a look. The documents should be passed directly to the estimating department. The COEP states that they should be inspected by the person who will be responsible for preparing the estimate. This should be the head of department – the individual who takes the responsibility for the estimate – not the estimator, who will later be appointed to deal with the task. It is most important that the early inspection is carried out by a person experienced in current procedures and documentation, and capable in decision making and effective in communicating with others.

The documents should first be checked that they accord with those listed in the letter of invitation, normally:

1. Two copies of the bills of quantities.
2. Two copies of the general arrangement drawings.
3. Two copies of the form of tender.
4. Addressed envelope for the return of tender (and priced bills if applicable).

If the documentation is not complete the fact should be reported immediately by telephone. If reference is made in the letter that certain sections of the bills will follow shortly and the tendering time is as stated in the preliminary invitation, an appeal should be made for a revision of time to comply with the CPSSST. This clearly states 'that the time for tendering should be calculated from the date of issue of the last section'.

The preliminaries sections of the bills need to be examined carefully at this stage, particularly the general and contractual particulars called for under A10-A37 SMM7. The drawings from which the bills were prepared should be listed in accordance with A11 and the drawings set out in General Rule 5 should be enclosed. Drawing number references should match those recorded, for example if the bills state drawing No. 90/3/2910C: if 'D' is supplied, then it must be questioned.

If further information is needed (to find the extent of temporary works, for example) more drawings may be sought. It is important that the estimator works from actual full-scale prints rather than trying to take in the information and making tracings at the consultant's office. As a general rule, clients should issue all information which is relevant and available at the tender stage.

The contractor may already have a guide to the value of the project. If not, he could get a rough guide to the tender figure by applying approximate rates to the principal quantities. The initial inspection of the tender documents is completed by producing a tender information (enquiry record) form which is similar to the Preliminary Enquiry form but with more details of the estimated cost and contract details. The COEP provides a form for both purposes called the Preliminary/ Tender Enquiry Form. This form is a valuable source of information because it provides management with a summary of the tender which is being prepared and can be kept for all previous tenders, whether successful or not.

Project appreciation 7

Introduction

Following management's decision to tender, the tender documents are given to the estimator to prepare the estimate. He should read the documents to gain an overall understanding of the project. A decision can then be made about the help needed from other departments for planning, procurement and commercial appraisal.

If a bill of quantities is available, enquiry schedules can be drawn up immediately, and documents will be prepared for suppliers and sub-contractors (see Chapter 8). Enquiries need to be sent promptly so that specialists have enough time to prepare their quotations.

Once the enquiries are under way, the estimator will broaden his understanding of the project by scheduling principal quantities and PC and provisional sums; he will undertake visits to site and if necessary the offices of the consultants.

Estimate timetable

For most tenders there is an absolute requirement to meet the submission date. The estimator must programme the activities needed to produce a tender to show how the deadline can be met and explain to other members of the team their part in the plan. Each job is different, and some dates such as those for the return of quotations require firm action to maintain the programme dates.

Time allowed for tendering is usually limited by the client's needs to start a project quickly. If the design stage has been delayed it is often the tender stage which is shortened. Flexibility is needed to concentrate on the critical parts of the estimate preparation. The estimator can press on to complete his work with a day or so to spare for reconciling and checking the estimate. Much of the early part of the tender period is given over to the dispatch of enquiries to suppliers and sub-contractors and setting up job files when a computer system is used. Figure

CB CONSTRUCTION — ESTIMATE TIMETABLE

Project : Lifeboat station
Ref No: T384 Date : 14/6/94

	June														July						
	13	14	15	16	17	20	21	22	23	24	27	28	29	30	1	4	5	6	7	8	11
Documents received	e																				
Decision making	e																				
Study documents		e																			
Mark up enquiries		e	e																		
Dispatch enquiries			b	b	b																
Date for return of mat prices											▓										
Date for return of s/c prices																▓					
Computer entry			a	a	a	a	a														
Visit site				e																	
Visit consultants					e																
Study methods, temp works and programme						ep	p	p	p	p											
Main pricing								e	e	e	e	e	e	e	e	e					
Extend bill rates and chase quotations												a	a	a	a	a	a				
Preliminaries																e	e	e	e		
Summaries and report																e	e	e	e		
Adjudication																				▓	
Prepare docs for submission																				▓	▓
Submission																					▓

Key : e = estimator, b = buyer, p = planner, a = estimating assistant

Fig 7.1 *A typical estimate timetable*

7.1 shows a simple timetable for producing an estimate and tender. This programme is simple to produce using a 'blank' standard form because many activities are common to all tenders.

The estimator is responsible for preparing the estimating timetable showing the key dates for him and the other members of the estimating team. As the team leader, the estimator will need to coordinate the other people in the team. The COEP gives a typical checklist for a coordination meeting (chaired by the chief estimator), which is presumably for large-scale or complex projects.

Schedules

The estimator should list all the prime cost and provisional sums to identify the work which will be carried out by other contractors. A summary of costs written into the bill will become part of the estimator's report for management at the adjudication stage. The example given in Fig. 7.2 shows that the structural steelwork and electrical sub-contractors will be chosen by the architect and no enquiries will be sent by the contractor at tender stage.

The summary of PC and provisional sums can also be used to show the attendances required by each nominated sub-contractor. SMM7 (A51.1.3) gives a list of the items of special attendance which must be given in a bill of quantities if required. The summary can include these items in the form of a checklist. Most of the costs of providing these attendances are evaluated when pricing the preliminaries because they can be considered in relation to the project as a whole.

The estimator needs to abstract the work which will be carried out by the main contractor, such as excavation, concrete work, brickwork and drainage. The trade abstract shown in Fig. 7.3 brings together all the pages to be priced under each trade heading, and helps the estimator to assign pricing duties when more than one estimator is working on the tender.

The estimating team

The roles of the members of the estimating team (Fig. 7.4) will vary from company to company and will depend on the size of the job. Some companies prefer to hand a copy of the documents to the buyer for sending out enquiries and others elect to keep control of this activity in the estimating section. A compromise would be for the estimator to abstract the materials and sub-contract packages, forming part of the estimate, and ask the buyer to select suitable companies to receive enquiries and coordinate and prepare documents for

CB CONSTRUCTION LIMITED	PC & Provisional Sums	Project	LIFEBOAT STATION		
		Ref.No: T384		Date: 14.6.94	

Bill Ref:	Description	Prov Sums	Prime Cost Sums			
			Gross	Discount	Nett	Notes for pricing preliminaries
	PC SUMS					**Special attendances**
6/1a	STRUCTURAL STEELWORK (STEELBUILD LTD)		23 000	575	22 425	GOOD ACCESS ROADS & HARDSTANDING
6/1e	ELECTRICAL INSTALLATION (NAME NOT GIVEN)		15 600	390	15 210	SCAFFOLDING COVERED STORAGE
6/2a	FIRE DOORS (NOM SUPP)		3 840	192	3 648	
	PROVISIONAL SUMS					**Prelims for defined prov sums :**
6/2m	CONTINGENCIES	5000				
6/2n	DRAINAGE TO SUMP	1000				
6/2p	GLAZED ROOF OVER ENTRANCE	3500				SCAFFOLDING PROTECTION & CLEANING
6/3	DAYWORK LABOUR	1000				
	ADD 110%	1100				
	DAYWORK MATERIALS	500				
	ADD 15%	75				
	DAYWORK PLANT	500				
	ADD 10%	50				
	Totals :	£ 12725	42 440	1157	41 283	

Fig 7.2 *List of PC and Provisional Sums at project appreciation stage*

CB CONSTRUCTION LIMITED Trade abstract		Project :	Lifeboat Station		
		Ref. No:	T384	Date :	22.6.94

Trade	Bill pages	Spec pages	Estimator
Earthworks	3/1–5, 4/35–37	2/1–12	JPM
Drainage	5/1–34	2/56–65	JPM
Concrete work	3/5–7, 4/38–40	2/13–15	JPM
Reinforcement	3/11, 4/43	2/15–16	JPM
Formwork	3/7–10, 4/40–42	2/17–22	JPM
Concrete sundries	3/7,11–13, 4/44,45	2/14	JPM
Precast concrete	4/46	2/35	PC
Brickwork	3/19–22	2/24–28	PC
Blockwork	3/20	2/27–32	PC
Brick sundries	3/22,23	2/24	PC
Timber	3/28–31	2/39–44	PC
Joinery	3/31–38	2/39–46	PC
Metalwork	3/29,39	2/47–48	PC
BWIC	3/55		JPM
Attendances	6/1		JPM

Fig 7.3 *Trade abstract for sections to be priced by the contractor*

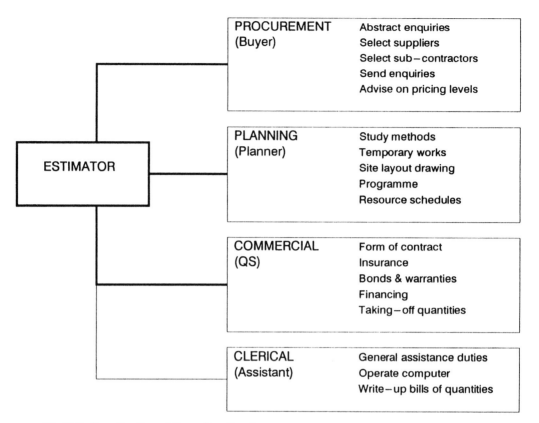

Fig 7.4 *Coordination of the estimating team*

despatch. This allows the estimator to keep control over what prices are sought, and the buyer can use his experience to get better prices. The commercial manager, or quantity surveyor, should be given the opportunity to comment on the form of contract proposed, insurances, bonds and financing requirements. If there are any onerous or unusual conditions which may cause problems with an unqualified bid, they should be challenged by writing to the quantity surveyor for a ruling. This should produce a more satisfactory result than leaving it for management to decide at the adjudication meeting when it is too late for an amendment to be made to the tender documents.

The planning engineer has a great deal to contribute. Instead of looking at the project from a financial point of view he will start by examining the layout of the site, methods for construction, temporary works requirements, distribution patterns, sequence of work, a preliminary construction programme and resource levels. A civil engineering estimator will usually assess the temporary works and programme the works himself because this is often the preparation needed for operational estimating techniques. A building estimator will tend to rate the items in a bill of quantities and rely on the planner for a more detailed examination of methods.

Visits to consultants and site

An estimator needs to examine the documents carefully before leaving his office. It may be helpful to mark up a site plan to highlight the main elements, such as: areas of scaffolding, access routes, existing and proposed services, fencing and external hard landscaping. A preliminary assessment can be made to find areas for general facilities such as site accommodation, cranage, storage areas and hardstandings. If any relevant information is missing then it might be questioned at the consultant's office or during the site visit.

In a perfect world there should be no need to visit the consultant's office – the tender documents should define the basis of an agreement to construct. There are, however, some benefits, as follows:

1. A critical assessment can be made of the progress made with the construction drawings, which would be important if there is a need for a quick start.
2. The estimator can seek clarification of layouts and details which were not included with the tender documents.
3. The names of adjoining property owners can be used to find sites for tipping and search for sources of fill.
4. The contractor can consider opportunities to offer an alternative bid in terms of time or design.
5. A contractor can explain its track record and interest in future projects.

A site visit must be made before a tender is submitted and should, wherever possible, be carried out by the estimator in person. The COEP provides a comprehensive standard report form which will cater for most projects.

Site features and existing buildings can be recorded on film. This will not only remind the estimator of site details during the pricing stage but will also prove an excellent means of communication at the adjudication meeting. Occasionally permission will be required to take photographs, such as: inside existing buildings, near sensitive production processes and of existing property where security must be preserved. Permission is not needed for photographs taken of any property from outside the site boundary.

In assessing the site, note should be made of the topography, whether it is situated on a hill or in a valley, for example. Some plants could signal that although the ground is dry in summer the same ground may be waterlogged in winter. The general levels of the site need to be related to the information gained from the site-investigation report, sometimes the remains of trial holes can be inspected. The contractor must assess the effect of a change of season from summer to winter. Some of the best information is available from residents and land owners. They often have valuable knowledge about flooding and ground conditions met during previous excavation work.

The site location and access routes must be examined. Urban sites need particular consideration. In built-up areas where cranes are to be used permission may be needed to encroach on air space; loading gantries, hoardings, protected walkways and temporary footpaths all need careful thought when pricing preliminaries. In country districts there could be a problem with the loads allowed in narrow lanes. Signboards may be needed to guide traffic from the nearest main road. Will crossovers be required from public roads to the site, or will they need to be constructed?

There are some questions about existing utility services which must be answered. Who supplies the services, where are the nearest connections and what sizes and capacities are available? Much effort can be saved if site toilets can be connected to a foul drain. Water will be needed early on when the site facilities are set up. Electricity supplies may be present but are they sufficient to meet the heavy demand of running plant? If an electrically powered tower crane is planned a large-capacity supply may be needed. The electricity company is usually reluctant to bring the new main in at an early stage unless the contractor is prepared to contribute to the capital cost.

Security is an increasing problem which must be checked during the site visit. Existing boundaries need to be related to the site boundary and compared with the fencing measured in the bill or described in the preliminaries. A short discussion with a resident might highlight the risks.

Finally and most important are the items which can only be priced by assessing the extent of the work on site. These include: demolitions (usually given as items in the tender documents with overall dimensions), alterations (given as spot items the complexity can be gauged in work-hours), and clearing site vegetation (measured in square metres and varies from site to site). Where the work is predominantly described in spot items, the estimator will make extended and often repeated visits to site on occasions meeting sub-contractors.

8 Enquiries to suppliers and sub-contractors

Introduction

Contractors have developed procedures to ensure that tenders are based on up-to-date prices for materials and specialist services. As the COEP states, 'The contractor's success in obtaining a contract can depend upon the quality of the quotations received for materials, plant and items to be sub-contracted'. To illustrate the point, the following breakdown has been taken from a typical building estimate, and shows that materials and sub-contractors account for 72% of the value of direct works:

Direct works analysis

Direct work – Labour	23%
Direct work – Plant	5%
Direct work – Materials	28%⎱
Domestic sub-contractors	44%⎰

Enquiries must be sent promptly but not at the expense of accuracy. Wrong or incomplete information will lead to delay in receiving comparable quotations. An orderly presentation will do much to avoid mistakes. The COEP provides some model forms for use in abstracting and suggests ways of selecting suitable suppliers and sub-contractors. These forms are further developed according to each company's procedures. Figure 8.1 shows a form which can be used for abstracting materials or sub-contract packages and later helps to record the receipt of quotations by highlighting the names of firms which have responded.

Most enquiries consist of photocopies of the relevant pages from the bills of quantities with a letter and any relevant drawings. Ideally those bill items which are not to be priced should be crossed out to avoid confusion. Although sub-contractors should be sent two copies of the bill pages which are to be

| CB CONSTRUCTION | ENQUIRY ABSTRACT | | [✓] materials | | | Project | Lifeboat Station | | |
| | | | [ ] sub–contractors | | | Ref. No. | T384 | Date: | 14/6/94 |

Ref.	Description	Approx. Quants	Bill pages	Spec pages	Drgs	Names	Telephone
M1	AGGREGATES						
	DOT TYPE 1	580T		2/5,6		LITTLEGREEN QUARRIES	
	DUST	55T				J.P.HEPPLE	
	20MM SINGLE SIZE	110T				SWANFIELD STONE CO	
M2	CONCRETE					SWANFIELD STONE CO	
						PINTO CONCRETE LTD	
	C15P	25m3		2/11–15		DRAY CONC SERVICES	
	C20	40m3					
	C35	400m3					
M3	REINFORCEMENT					BARBEND FABRICATION	
						DOWLAIS STEEL	
	A393	1240M2		2/19,20		OAKFORD REINFORCEMENT	
	10MM HY	1.1T					
	12MM HY	3.8T					
	16MM HY	5.6T					
M4	PRECAST CONC					TELGAN CONC PRODUCTS	
						HILBERG CASTINGS	
	LINTELS	27NR	3/22	2/22			
	KERBS	360M	4/46				
M5	BRICKS/BLOCKS					SIMGROVE BRICK	
						HEPPLE BRICK CO	
	FACINGS	11000NR	3/19–22	2/28–30		BUSH BROS LTD	
	100MM BLOCKS	550M2					
	140MM BLOCKS	200M2					
M6	JOINERY					GOTHIC JOINERY LTD	
	AS BILL PAGES		3/31–37	2/36–40	SK1,2,3	ST.ANNES TIMBER	
					B/27/204	SHIRE MANUFACTURING	

Fig 8.1 *Abstract form for materials or sub-contract enquiries*

returned, this practice has died out largely because sub-contractors prefer to photocopy their priced bills. With the introduction of CPI and SMM7, bill descriptions now have references to the specification clause which apply to the item of work. This helps the estimator to ensure that he sends the correct specification clauses with the enquiries.

In the past estimators and buyers have kept lists of companies who can supply materials and services in a card index of names usually in trade order. Since desktop computers have been introduced with user-definable databases, many contractors maintain name and address files which can be searched for trade contractors within travelling distance of the site. The software is usually able to address the letters and envelopes and keep track of previous performance, although feedback from site on a sub-contractor's behaviour is not always consistent or reliable.

When tendering in an unfamiliar location, the estimator must allow time when visiting the site to tour the area, and note the volume of work going on and what stage individual jobs have reached. Site hoardings which list the sub-contractors being used in an area are a valuable source of information about specialists who are acceptable to your competitors. The local library is a useful source for obtaining names, as is the local newspaper. The estimator will also refer to nationally published directories and the *Yellow Pages*. If new sub-contractors are to be invited to tender the buyer should telephone first to ascertain their ability and willingness to prepare a quotation.

Enquiries for materials

Many construction organizations have a standard form of enquiry for suppliers, and a typical example is given in Fig. 8.2. Enquiries should give the following information:

1. Title and location of the work; some suppliers complain that enquiries from different contractors are received in varying formats, often making it impossible to decide whether they are for the same job;
2. Description of the materials, supported by specifications;
3. Approximate quantities, so that bulk discounts can be quoted;
4. Date by which the quotation is needed; 7 to 10 days would seem reasonable although those with complex fabrication work to price (joinery suppliers, for example) may need longer; the most successful approach where time is limited is to say so, and request cooperation by responding as soon as possible;
5. Name of the estimator dealing with the tender;
6. The contract period with a guide to the dates for deliveries; it can only be a

Our ref : T384/M5   16th June 1994

Simgrove Brick Ltd
Unit 3, Northbridge Industrial Estate
Northbridge
NB3 5MGG

Dear Sirs,

NEW LIFEBOAT STATION
BEACH LANE, STANSFORD

We are tendering for the above project and ask you to submit your best rates
for the following items to be delivered to the site. The project is due to start
on 2nd September 1994. If you have any queries in relation to this enquiry
please contact our Regional Buyer, Mr. Frank Applecourt.

Please reply by 28th June 1994 and state any known or anticipated price rises
likely to affect our tender.

Yours faithfully,

Item	Description	Approx quants	Supporting documents enclosed
1.	Facing bricks	11,000	Bill pages 3/19 – 22 Spec pages 2/28 – 30
2.	100mm blocks	550m²	
3.	140mm blocks	200m²	

Fig 8.2 *A typical enquiry for materials*

guide because the start date is rarely known and the construction programme is not yet available;

7. Whether firm price or fluctuating price;
8. Minimum discount terms required;
9. Any limitations on access to the site.

The supplier should make every effort to meet the specified submission date and tell the contractor if a delay is expected. If the supplier is himself awaiting information from his sources, he should submit a quotation with a clear statement where prices are to follow.

There is not much point in keeping price-lists unless they are up-dated regularly. It is more important for an estimator to compile a library on material characteristics, quality, sizes and performance standards. A knowledge of available materials and products will enable the estimator to consider alternatives which comply with the specification.

Enquiries to sub-contractors

There are three kinds of domestic sub-contractor who will be approached at tender stage. The conventional sub-contractor who provides a complete service, labour-only sub-contractors who are supplied with their materials and plant, and labour and plant sub-contractors who receive their materials from the main contractor. Sub-letting work to specialists is an attractive arrangement for contractors because much of the technical and financial risk is passed to another party, and a profit is almost guaranteed (providing the work goes to plan). On the other hand, sub-contractors can benefit from increased quantities and certain variations. Most contracts state that the main contractor shall get written permission before sub-letting any of the work, and in some (now rare) instances employers will not allow the use of labour-only sub-contractors.

Domestic sub-contractors will need a lot of information about the site, contract conditions, programme, the specification and extent of work. The example given in Fig. 8.3 shows a standard enquiry letter which can be stored as a word-processor file and tailored for each contract and trade. This example is lacking guidance on the timing of the work, probably because this can be found in the extract from the preliminaries. On larger projects, an outline programme might be available and the sub-contractor will be asked to prepare his own tender programme with information about extended delivery periods which might affect the progress of the works.

The COEP lists the details to be given in a contractor's enquiry letter, as follows:

16th June 1994

Dear Sirs,

<u>NEW LIFEBOAT STATION,</u>
<u>BEACH LANE, STANSFORD</u>

We invite you to tender for the PAINTING work for a proposed Lifeboat Station and enclose the following details which describe the quality and quantity of the work:

Preliminaries pages:	1/2-12	
Bill pages:	3/45-48	(2 copies)
Specification pages:	2/39-40	
Drawings:	D/206/1	
Form of Tender		(2 copies)

The names of the parties, general description of the works, and details of the main contract are given in the extract from the preliminaries. Your form of tender, priced bill of quantities and daywork rates must be delivered to this address to arrive by <u>4th July 1994.</u>

The form of sub-contract will be DOM/1 incorporating all relevant published amendments and the following:

Payments	:	Monthly
Discount to main contractor	:	2.5%
Fluctuations	:	Firm Price
Liquidated damages	:	£1200 per week
Basis of daywork	:	current RICS definition
Retention	:	5%
Method of Measurement	:	SMM7
Defect liability period	:	six months

We will provide all sub-contractors with water, lighting and electricity services near the work and common welfare facilities on site. Sub-contractors will be required to provide the following services and facilities:

a.   unloading, storing and taking materials to working areas
b.   power and fuel charges to temporary site accommodation
c.   clearing-up, removing and depositing in designated collection points on site all rubbish or other surplus or packing materials
d.   temporary accommodation and telephones
e.   day-to-day setting out from main contractor's base lines

If you have any queries about this enquiry please contact the estimator for the project, Mr John Edwards.

Would you please confirm by return your willingness to tender by the date for tender.

Yours faithfully,

Fig 8.3 *Sample enquiry letter to a domestic sub-contractor*

1. Site address and location (with a map if necessary).
2. Name of employer, and professional team.
3. Relevant details of main and sub-contract.
4. Any amendments to the standard conditions, including bonds and insurances.
5. A request for daywork rates.
6. Date for return of quotations.
7. General description of works.
8. Details of access, site plant and other facilities available.
9. Where full contract details and drawings can be inspected.
10. Contract period, programme and any phasing requirements.
11. Any discounts to be included.
12. Two copies of the relevant sections of the bills of quantities.
13. Copies of drawings and schedules where applicable.
14. Services and attendances to be provided by the main contractor.
15. A clear statement of how fluctuations will be dealt with.

Labour-only sub-contractors usually quote under different arrangements; in particular they often expect:

1. Setting out by the main contractor.
2. Weekly or fortnightly payments by the main contractor.
3. Modified retention sums to reflect the extent of their work.
4. Materials delivered, unloaded and sometimes taken close to the point of fixing.
5. Major items of plant (which will be used by other sub-contractors) to be provided by the main contractor.

Although most enquiries are in the form of photocopies of bill pages, it is important that the estimator clearly states the portions of the work which the labour-only sub-contractor is expected to carry out. As an example, a concrete specialist might be asked to price placing concrete, fixing reinforcement and labour and materials in fixing formwork. Another firm might be asked to lay concrete in floor slabs and power float the surface.

The Building Employers Confederation produced in 1985 a 'Code of Procedure and Checklists for the Letting and Management of Domestic Sub-contract Works' which recommends a tendering procedure which mirrors that suggested by the NJCC for main contracts. In other words, sub-contractors should be asked for their willingness to tender, there should be full tender documents, sufficient time must be given for preparing tenders, and sub-contractors should be told about their performance.

Tender planning and method statements

Introduction

The estimating team will consider construction methods and employ planning techniques to:

1. Highlight any critical or unusual activities.
2. Examine alternative ways of tackling the work.
3. Calculate optimum durations for temporary works and plant.
4. Reconcile the labour costs in the estimate with a programme showing resources.
5. Determine the general items and facilities priced in the preliminaries section of the bill.
6. Check whether the time for completion is acceptable.

The effort needed will depend on the size and complexity of the project, the proposed use of heavy plant and the design of major temporary works. Estimating for civil engineering work in particular is dependent on an examination of alternative methods and pre-tender programmes. A civil engineering estimator usually produces a resourced programme to price the work operationally.

Pre-tender programmes are prepared by either the estimator or planning engineer, or by working together. The choice depends on company policy, size of project and type of work. The planning engineer's contribution can be seen as producing an appraisal of labour and plant resources and general items – in other words, the estimator expresses his solutions in terms of cash, the programmer deals with time. The aim is to reconcile one with the other.

The role of the planning engineer

In a competitive market it is important to look for ways to construct the project more economically. Applying planning techniques can have opposite consequences – increasing the value of the tender when problems are identified and reducing the estimate when methods can be adopted which reduce individual and overall durations. The team must, however, look for the solution which reflects the 'true' value of construction. The role of the planning engineer is wider than just producing a programme. His input to a tender can also include:

1. Producing site layout drawings, which are used to locate temporary facilities, such as concrete batching plant, cranage, access routes, restrictions, areas for accommodation and storage, location of services, overhead service, temporary spoil heaps, and areas which will need reinstatement;
2. Examining the most suitable methods in relation to the design and the temporary works required;
3. Preparing method statements not only for pricing purposes but also for submission to consultants when requested;
4. Producing cashflow forecast charts for management and clients who need them;
5. Providing staff structure and resource histograms for general labour and plant.

The planning engineer will often have a better understanding of current site practice and will be better placed to collect data from monitoring exercises on site. His experience of completed work will be important especially where the overall duration of a project could be reduced. Shorter contract periods can have a substantial effect on the cost of preliminaries where time-related costs (mainly staff, cranage and scaffolding) account for as much as 12–20% of a tender figure.

Method statements

Method statements are written descriptions of how items of work will be carried out. They usually deal with the use of labour and plant in terms of types, gang sizes and expected outputs.

There are many reasons why method statements are prepared during the tender stage. It is unlikely that an estimator will prepare a written method statement for his own use but if any of the following requirements exist then he will commit his thoughts to paper:

1. The client's advisers may ask for a method statement to accompany the tender, to satisfy themselves that the contractor has an understanding of the difficulties and has considered suitable ways of overcoming them.

2. The quality management scheme adopted by the organization may entail method statements for work worth more than a certain value.
3. Management contractors usually ask for method statements because the works contractor may be proposing working methods which may lead to interface problems with other specialists on the site.
4. In satisfying the need for safe systems of work, an estimator might develop a method statement with a demolition contractor, for example, before agreeing a price to be incorporated into the tender.
5. Large-scale activities needing a combination of items of plant and labour are difficult to price on a unit rate basis and cannot be started without an examination of methods and resources.
6. Where the estimator has investigated an alternative design he will need to assess the effect these changes will have on other elements of the construction.
7. Part of the handover information prepared for successful tenders is a description of the assumptions made by the estimator.

Many contractors are reluctant to submit a detailed method statement at tender stage because their ideas could be used by other parties without any financial return. A preliminary document can be prepared (see Fig. 9.1) based on the broad assumptions made at tender and is likely to include extracts from the company's manuals for safety and quality management. This method statement can also be of benefit to the contractor because it is an ideal vehicle for:

1. Qualifying the tender.
2. Identifying dates when information is required from the client or his advisers.
3. Indicating when instructions are required for dealing with nominated sub-contractors and provisional sums.
4. Explaining the limitations of temporary works; a contractor might have allowed for earthwork support but not sheet piling, for example.

Tender programmes

The tender programme will fix an overall time for the project, from which the estimator can determine times for sections of work in main stages such as:

1. Design and mobilization.
2. Substructure.
3. Independent structures.
4. Superstructure.
5. Internal trades, finishes and services.
6. External works.

101

Outline Method Statement
for

New Offices
for

Fast Transport PLC
North Lane, Stansford

Site Location   At the site of the existing transport yard of
                    Fast Transport PLC, North Lane, Stansford.
                    Access will be through the main entrance
                    gate.

Restrictions/Access Incoming traffic will be directed to use the
                    north access road and will leave the site
                    along the road next to the canteen.

                    The live oil and gas mains will be protected
                    during the contract period and the fibre
                    optics cable will be carefully exposed by
                    hand dig and protected in accordance with the
                    specification prior to piling equipment
                    entering the site.

Sequence of work   Our tender programme T354/P1 shows the
                    preferred sequence of activities. The aim is
                    to start at the east end of the building
                    progressing to the west. The site will be
                    filled with a stone layer on a ground
                    improvement mat immediately after the site is
                    levelled.

                    Concrete floors will be started after the
                    columns have been cast and before the upper
                    floors are constructed. External paving will
                    be carried out in the last quarter of the
                    contract period. The drain connecting manhole
                    3 to the existing foul sewer will be
                    completed early to provide disposal from
                    temporary facilities.

Design development

Detailed drawings of the roof cladding will be produced by our specialist sub-contractor. These will clarify the scope of the work giving fixing details, sequencing and weathering procedures. Roof flashings will not be made until formal approval has been received by our sub-contractor.

Temporary works

An independent scaffold will be erected to each external face of the building, and a mechanical hoist will be provided near the north-west corner.

Safety

Anyone working on or visiting the site will be required to wear safety helmets and operatives will use other protective clothing depending on the type and location of work. The sides of the drain trench next to the oil tank will be supported with trench sheeting and we will provide barriers next to all excavations where a danger exists. The agent will attend regular meetings with the client's safety officer and cooperate with site regulations to maintain the client's good safety record.

The safety performance of the site is monitored by line management who report to regular safety audit meetings; and external consultants inspect our compliance with current legislation at intervals of no more three weeks.

Supervision

Our management structure for the area is shown in the diagram attached. We will adopt a flexible approach to site supervision, increasing the numbers in line with our programme requirements.

Quality plan

The site agent will be responsible for drawing up a quality plan for the project with assistance from the area planning engineer. The control and monitoring framework is given in the company's general procedures and QA manual.

Fig 9.1 *Example of a tender method statement for submission to a client*

Information about these periods is essential to the estimator, enabling him to calculate times for:

1. Staff requirements.
2. Site accommodation.
3. Mechanical plant and equipment.
4. Temporary works such as falsework and scaffolding.
5. Increased costs for firm price tenders.
6. Work affected by the seasonal weather changes such as drying out buildings, heating, protection and landscaping.

The overall section durations can be used to check workforce levels and items of plant such as excavators and cranes which often remain on site for continuous periods. There may be times of excessive demand for plant and labour which will call for a levelling exercise to balance resource needs.

The estimator will want specific information from the planner. He must be clear about what he needs from the programme so that the planner will concentrate on what is important.

To illustrate the point, an estimator has identified a clear run of brickwork comprising 250 000 facing bricks. He knows that the current price for laying is £220 per thousand and the all-in rate for bricklayers is £8.00 and labourers £6.50 per hour. The total cost of laying 250 000 bricks would be:

$$250\,000/1000 \times £220.00 = £55\,000$$

Assuming each bricklayer is serviced with half the time of a labourer, each bricklayer's effective rate is:

$$8.00 + (6.50/2) = £11.25$$

Therefore the total time included in the rates would be:

$$£55\,000/£11.25/\text{hour} = 4889 \text{ hours work}$$

The planner has decided to use an output of 50 bricks per hour.

$$250\,000 \text{ bricks at } 50 \text{ bricks/hour} = 5000 \text{ hours work}$$

This is clearly close to the estimate. Calculating gang sizes and number of gangs will not change the rate but will alter the cost of ancillary facilities such as scaffolding and mixers. Most activity durations can be derived from the product of quantities and standard outputs (see Fig. 9.2) but much of the tender will be based on offers received from specialist sub-contractors and labour-only sub-contractors. These firms will be asked to provide information about the time they will be on site and the effect of delivery periods on the main contractor's programme.

The tender programme must allow for recognized public/industry holidays, inclement weather and the peak summer holiday period which leads to a slowing of progress which may be reflected in output. Scaffolding, fixed cranes and

Activity	Quant	Output hrs/unit	Man wks	Gang size	Duration wks
1 Setting up and setting out					
2 Excavation and filling	1035m³	0.13	2.9	JCB+lab	3
3 Foundations formwork	820m²	1.20	21.9	4 carp	5
4 Foundations concrete	419m³	1.20	11.1	3 lab	4
5 Underslab drainage	178m	0.50	2.0	2 lab	2
6 Concrete ground floors	137m³	1.00	3.0	4 lab	1
7 Columns formwork	279m²	1.20	7.5	2 carp	4
8 Columns concrete	21m³	3.50	1.6	3 lab	1
9 Floors and beams formwork	962m²	1.60	34.2	4 carp	8
10 Floors and beams concrete	203m³	2.20	10.0	4 lab	3
11 External walls	1870m²	1.50	62.3	8 brklayer	8
12 Roof timbers	2300m	0.20	10.2	4 carp	3
13 Roof covering	685m²	0.50	7.6	4 carp	2
14 Windows	£25500 sub-contract				4
15 Services 1st fix	£43000 sub-contract				6
16 Plasterwork and partitions	£43500 sub-contract				8
17 Joinery	£34000 sub-contract				8
18 Ceilings	945m²	1.00	21.0	4 fixers	5
19 Services 2nd fix	£28000 sub-contract				5
20 Painting	3500m²	0.30	23.3	4 paint	6
21 Floor coverings	£12300 sub-contract				3
22 External work and drainage	£16700 labour		52.0	5 lab	10
PRELIMINARIES					
Agent	(24 plus 3 mobilization)				27
Engineer					17
Foreman					26
Crane					7
Forklift					10

Bar chart columns: 1994 (September 1 2 3 4 | October 5 6 7 | November 8 9 10 11 12 13 | December 14 15 16) 1995 (17 | January 18 19 20 21 | February 22 23 24 25 26 | March 27 28 29 30)

Fig 9.2 *Example of a tender programme*

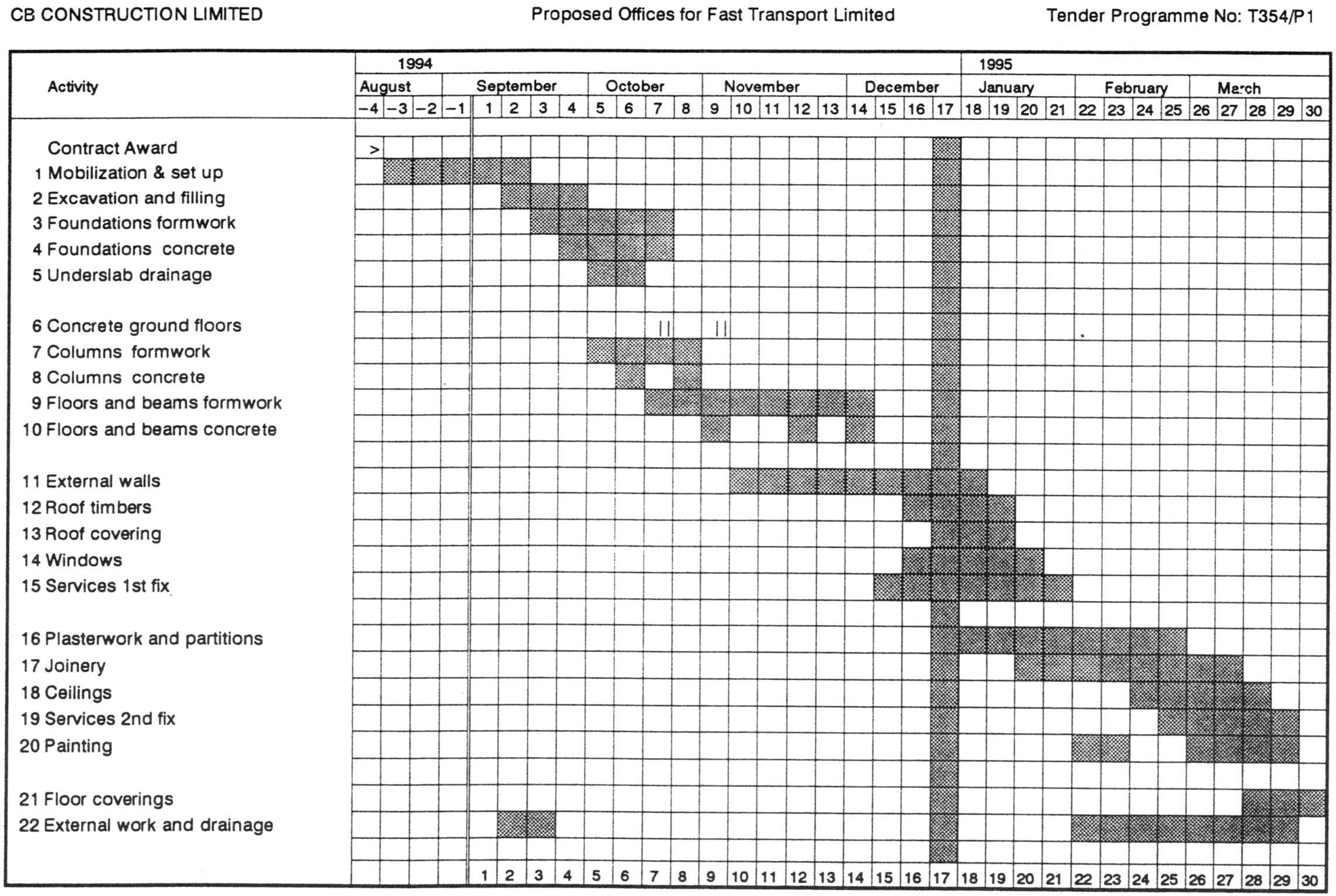

CB CONSTRUCTION LIMITED

Proposed Offices for Fast Transport Limited

Tender Programme No: T354/P1

Activity	1994 August			September				October					November					December				1995 January					February				March				
	-4	-3	-2	-1	1	2	3	4	5	6	7	8	9	10	11	12	13	14	15	16	17	18	19	20	21	22	23	24	25	26	27	28	29	30

Contract Award
1 Mobilization & set up
2 Excavation and filling
3 Foundations formwork
4 Foundations concrete
5 Underslab drainage
6 Concrete ground floors
7 Columns formwork
8 Columns concrete
9 Floors and beams formwork
10 Floors and beams concrete
11 External walls
12 Roof timbers
13 Roof covering
14 Windows
15 Services 1st fix
16 Plasterwork and partitions
17 Joinery
18 Ceilings
19 Services 2nd fix
20 Painting
21 Floor coverings
22 External work and drainage

Fig 9.3 Example of a programme submitted with a tender

supervision are items which will incur the largest costs during shutdown periods and must be included in the project overheads schedule.

When a client or his advisers request a programme at tender stage the contractor will submit a preliminary or outline programme, such as the example given in Fig. 9.3. The contractor is often unclear about the role of such a programme in vetting tenders. Sometimes contractors have used the opportunity to offer completion sooner than expected and thereby try to gain an advantage over the competition. The drawback is that if the project is delayed but still finishes within the original duration the contractor will have difficulty in recovering the costs of delay and disruption to the work.

Resource costs – labour, materials and plant

Introduction

There was a time when the unit costs of labour and plant were calculated from first principles; the assumption being that the company employed operatives in sufficient numbers to carry out the work and provided its own plant. A more realistic approach today would be to find the current market rates paid for labour near the site and look at the market prices for plant hire. This information is readily available as feedback from current jobs and plant hire rates can be obtained from plant specialists. Another change has come with computers. The importance of establishing accurate rates for labour, materials and plant, before pricing the bill of quantities, has reduced because computer programs allow the estimator to change unit rates for resources at any stage of the tender period.

Labour rates

A method for estimating all-in rates for labour is given in the CIOB Code of Estimating Practice (COEP). This has been adopted by many publications, professional bodies and contracting organizations as a reasonable basis for calculating the cost to employ an operative. The example given in the Code uses the formula:

Hourly rate = annual cost of employing an
operative/actual hours worked

During the first half of the 20th century, builders calculated labour rates by looking at weekly costs. This was a little easier to do but lacked the precision of the current method. The main reasons for calculating costs and hours on an annual basis are:

1. The effect of annual and public holidays on the number of hours for payment and hours of work.

2. Many of the costs are incurred for the weeks at work, and so exclude annual and public holidays.
3. Overtime working often depends on the proportion of summer and winter working, because longer working hours are available and used in the summer period.
4. The training levy is sometimes expressed as an annual lump sum for each operative.

The COEP calculation is clearly a theoretical approach which should be checked periodically against recorded costs. The main variance is commonly the amount paid for 'bonuses', such as attraction money, plus rates for semi-skilled operatives, spot bonuses and locally agreed payments.

The estimator needs to be aware of some of the difficulties associated with calculating labour costs, and should answer the following questions:

1. Are there enough skilled work people in the area? If not, will they need to be paid increased rates to work on the site or is there a need to import labour from outside the area?
2. How many operatives will be paid travelling expenses and will any key people receive a subsistence allowance?
3. Will there be any local union agreements which affect the wage levels, such as those found in the petrochemical industry?
4. Will bonus payments and enhanced wages be self-financing?

Some organizations, typically those which employ their own regular labour force, build up labour rates for every job. This allows changes to be made for the type of work, time of year and location. It must be said, however, that in recent years the nationally agreed wage rates have not reflected the rates paid in the marketplace. On the one hand, where skilled labour is scarce, labour costs more, and during times of recession labour rates fall. There is an argument that an estimator will price work quicker if a constant labour rate is used for several months. Global adjustments can always take place at the adjudication stage providing an analytical approach to pricing is used. Where computer databases are used, fine-tuning of the labour element can take place at any time before tender submission.

Figure 10.1 illustrates the all-in rate calculation using a spreadsheet model. Travelling and subsistence costs have been omitted on the assumption that they are better assessed when calculating the project overheads. Changes can be made to any of the figures and the following are the items which might change from job to job:

1. Time of year – the proportion of work carried out during 'summer' weeks;
2. Number of hours worked each week – the normal working hours are

CB CONSTRUCTION LTD	Fast Transport Ltd			1992–93	
Description		Entry col	Calc col		
SUMMER PERIOD	Number of weeks	30			
	Weekly hours	44			
	Total hours		1320		
	Days annual hols	14			
	Days public hols	5			
	Total hrs for hols		–167		
WINTER PERIOD	Number of weeks	22			
	Weekly hours	39			
	Total hours		858		
	Days annual hols	7			
	Days public hols	3			
	Total hrs for hols		–78		
SICKNESS	Number of days (say winter)	8	–62		
TOTAL HOURS FOR PAYMENT			1870		
%	Allowance for bad weather	2	37		
TOTAL PRODUCTIVE HOURS			1833	CRAFT	LAB
		craft	lab		
ANNUAL EARNINGS	Guar Min Wage	159.51	135.91		
	Attraction bonus	10.00	0.00		
	Total weekly rate	169.51	135.91		
	Hourly rate of pay (39th)	4.35	3.48		
	Annual earnings =	1870.4 x hourly rate		8136.24	6508.99
	Public holidays	67.4 hours x rate		293.19	234.55
ADDITIONAL COSTS	NON–PRODUCTIVE OVERTIME (time+half only)				
	Hours per week ..summer		2.50		
	Hours per week ..winter		0.00		
	Hours per year ..summer		65.50		
	Hours per year ..winter		0.00		
	Cost of non–prod overtime			267.90	228.26
	SICK PAY per day	12.1			
	for 5 days =			60.50	60.50
TRADE SUPERVISION					
	No.of tradesmen per foreman	8			
	Plus rate for foreman	0.25			
	% of time on supervision	50		537.74	436.04
WORKING RULE AGREEMENT					
	Tool money per hour	0.04			
	Plus rate .. per hour	0.12		299.26	224.45
	Sub–total			9594.83	7692.79
OVERHEADS	TRAINING LEVY %	0.25		23.99	19.23
	NATIONAL INS %	8.6		825.16	661.58
	HOLIDAYS WITH PAY	19.2		902.40	902.40
	Sub–total			11346.38	9276.01
SEVERANCE PAY and SUNDRIES	%..	2		226.93	185.52
	Sub–total			11573.31	9461.53
EMPL. LIABTY & 3rd PARTY INS . %		2		231.47	189.23
ANNUAL COST OF OPERATIVE				11804.78	9650.76
Divide by Total Productive Hours ..		1833			
	COST PER HOUR =			£6.44	£5.27

Fig 10.1 *Calculation of all-in rates for labour using spreadsheet software*

thirty-nine per week throughout the year, but in the summer more working hours can be achieved;

3. The allowance for bad weather – depends on time of year, exposure to the weather and height above sea level;

4. Attraction bonus – is the non-productive element needed to match the going rate for skilled and semi-skilled people;

5. Trade supervision – is rarely included in the all-in rate today because it is better to consider all aspects of site supervision while assessing project overheads;

6. Extra payments for special skills – the Working Rule Agreement specifies many additional payments (to be added to the labourer's rate) principally for plant operatives;

7. Employer's liability insurance – although related to the labour value may be part of a general assessment of liabilities in the project overheads schedule.

Spreadsheets are used for these repetitious calculations because various combinations can be tried out, hence the phrase 'what if calculations'. In Fig. 10.2, supervision and insurances have been removed so that they can be considered in pricing preliminaries. A longer working week is envisaged during a 35-week summer period. The overall effect is a reduced hourly rate.

For the analysis of rates throughout this book, the labour rates have been rounded off to £8.00/hour for craftsmen and £6.50/hour for labourers, the assumption being that the recession of the early 1990s will be replaced by gradual growth later in the decade.

Material rates

Quotations should be obtained for all materials, not only because prices can fluctuate unpredictably but also the haulage rates to various sites could be different depending on their distance from the supplier; and the size of loads can significantly affect the transport costs. The following factors are considered by the estimator in building up the material portion of a unit rate:

1. Check that the materials comply with the specification – the estimator may consider the use of an alternative product if it is cheaper and from experience this is a satisfactory choice which the contract administrator is likely to accept. A common example is the use of cement replacements and additives in ready-mixed concrete which ironically are readily accepted by the Department of Transport and water industry and sometimes rejected by architects for building work. Many specifications envisage the use of alternatives with

CB CONSTRUCTION LTD		Fast Transport Ltd		1992–93	
Description		Entry col	Calc col		
SUMMER PERIOD	Number of weeks	35			
	Weekly hours	50			
	Total hours		1750		
	Days annual hols	14			
	Days public hols	5			
	Total hrs for hols		−190		
WINTER PERIOD	Number of weeks	17			
	Weekly hours	39			
	Total hours		663		
	Days annual hols	7			
	Days public hols	3			
	Total hrs for hols		−78		
SICKNESS	Number of days (say winter)	8	−62		
TOTAL HOURS FOR PAYMENT			2083		
% Allowance for bad weather		2	42		
TOTAL PRODUCTIVE HOURS			2041	CRAFT	LAB
		craft	lab		
ANNUAL EARNINGS	Guar Min Wage	159.51	135.91		
	Attraction bonus	10.00	0.00		
	Total weekly rate	169.51	135.91		
	Hourly rate of pay (39th)	4.35	3.48		
Annual earnings =	2082.6 x hourly rate			9059.31	7247.45
	Public holidays 73.4 hours x rate			319.29	255.43
ADDITIONAL COSTS	NON–PRODUCTIVE OVERTIME (time + half only)				
	Hours per week ..summer		5.50		
	Hours per week ..winter		0.00		
	Hours per year ..summer		171.60		
	Hours per year ..winter		0.00		
	Cost of non–prod overtime			701.84	598.00
	SICK PAY per day	12.1			
	for 5 days =			60.50	60.50
TRADE SUPERVISION					
	No.of tradesmen per foreman	8			
	Plus rate for foreman	0.25			
	% of time on supervision	0		0.00	0.00
WORKING RULE AGREEMENT					
	Tool money per hour	0.04			
	Plus rate .. per hour	0.12		333.22	249.91
	Sub–total			10474.16	8411.29
OVERHEADS	TRAINING LEVY %	0.25		26.19	21.03
	NATIONAL INS %	8.6		900.78	723.37
	HOLIDAYS WITH PAY	19.2		902.40	902.40
	Sub–total			12303.53	10058.09
SEVERANCE PAY and SUNDRIES	%..	2		246.07	201.16
	Sub–total			12549.60	10259.25
EMPL. LIABTY & 3rd PARTY INS . %		0		0.00	0.00
ANNUAL COST OF OPERATIVE				12549.60	10259.25
Divide by Total Productive Hours ..		2041			
COST PER HOUR =				£6.15	£5.03

Fig 10.2 *Calculation of all-in rates for site working 50 hours/week with an extended summer period of 35 weeks and supervision and insurance priced in preliminaries*

the statement 'subject to the approval of the Contract Administrator', for example.

2. The supplier may want payments for the costs of transport or small load charges. Ready-mixed concrete suppliers, for example, impose extra payments for part loads. The cost can be significant and must be considered where small concrete pours are expected.

3. Some products are manufactured in fixed sizes which are the minimum that can be ordered. An estimator may have received a price of £2.35 per metre for polythene pipe for a job which needs only 15 m. If the minimum coil size is 30 m then the estimator must consider the likelihood of using the pipe on another site which might involve a storage cost. Alternatively, it might be more realistic to allow £4.70/metre (including waste) in this tender.

4. The quantity required for each unit of work must be considered for each material. Estimators should keep a note of the conversion factors they need for commonly used materials. For example, a half brick wall has 60 bricks per square metre, 2.1 tonnes of stone may be needed for each cubic metre of hardcore, and 0.07 litre of emulsion paint might be the coverage for work to plastered ceilings.

5. Unloading and distributing materials are activities which can be priced in the unit rate calculation or dealt with as a general site facility in the project overheads. Often a combination of both is needed. With facing bricks, for example, the price for bricks will include the cost of mechanical off-loading; whereas distributing bricks around the site could be catered for by including a forklift in the project overhead schedule.

6. If the specification, or preliminaries clauses, call for samples of certain materials the estimator needs to ascertain the cost. Usually a supplier will provide samples without charge. Testing of materials, on the other hand, is usually undertaken by an independent organization, and as such must be specified or preferably included as an item in the bill of quantities. The cost of testing will be assessed when the overheads are calculated.

7. An allowance for *waste* is difficult to estimate. The standard methods of measurement state that work is measured net as fixed in position (SMM7 3.3.1) and the contractor is to allow for any waste and square cutting (SMM7 4.6 e and f) and overlapping of materials (Fabric reinforcement E30.M4, for example). CESMM3 Section 5 states that the quantities shall be calculated net using dimensions from the drawings and that no allowance shall be made for bulking, shrinkage or waste. The questions which the estimator must consider are: is there a selection process needed on site to achieve the quality specified (such as selecting facing bricks to produce a specific pattern)? Are the materials likely to suffer damage in the off-loading and handling stages? Is the design going to lead to losses in cutting standard components to fit the site dimensions? Is the site secure from theft and vandalism? Will the finished

work be protected from damage by following trades? Has the company had previous experience with the materials? Will some materials be used for the wrong purpose, such as using facing bricks below ground level to avoid ordering a few cheaper bricks?

An estimator will need help in making these decisions. Guidance can be found in price books or research papers and the company should collect information from previous projects.

Plant rates

The plant supply industry can provide a wide range of equipment throughout the United Kingdom. It can offer hire or outright purchase, and in some cases lease and contract rental schemes. The following steps can be taken at tender stage to assess the mechanical plant to be used:

Step 1   Identify specific items of plant needed by looking at quantities and methods. The machine capacities can be found by assessing the rates of production required. Examine the tender programme for overall durations.

Step 2   Obtain prices; the *sources of plant* are:
(a)   Purchase for the contract
(b)   Company-owned plant
(c)   Hire from external source.
In practice the *sources of prices* are:
• Calculate from first principles
• Internal plant department rates
• Hirers' quotations
• Published schedules.

Step 3   Compare plant quotations on an equal basis perhaps by using a standard form (the Code of Estimating Practice provides a typical Plant Quotations Register).

Step 4   Calculate the all-in hourly rate for each item of mechanical plant (the Code of Estimating Practice gives a calculation sheet). The main parts of the calculation are:
1.   Cost of machine per hour (including depreciation, maintenance, insurances, licences and overheads).
2.   All-in rate for operator (the operator may work longer hours than the plant because of the time needed for minor repairs, oiling and greasing; the National Working Rule Agreement suggests how

115

much time should be added to each eight-hour shift; it also lists extra payments for continuous extra skill or responsibility in driving various items of plant).

3. Fuel and lubricants (the amounts of fuel consumed will depend on the types and sizes of plant; the average consumption during the plant life is used).

4. Sundry consumables (where, for example, the plant specialist is unable to accept the risk of tyre replacement on a difficult site or any costs beyond 'fair wear and tear').

The cost of bringing plant to site is usually calculated in project overheads (preliminaries) when the transport of all plant and equipment is considered.

Step 5   Decide where to price plant – either in the unit rate against each item of measured work or in the project overheads. This decision might be made for the estimator if the company's procedures dictate the pricing method. Plant which serves several trades should be included in the project overheads, such as cranes, hoists, concrete mixers, and materials handling equipment. Estimators also price the erection of fixed plant in the project overheads together with the costs of dismantling plant on completion.

Unit rate pricing

Introduction

The estimator must press on with the pricing stage without delay and cannot afford to wait for the quotations from sub-contractors and suppliers. Once the basic rates have been calculated for labour and plant, pricing notes can be written for work which will not be sub-let; such as placing concrete, alterations and brickwork. The pricing form in Fig. 11.1 shows an estimator's notes for fixing ironmongery with spaces for the prices from suppliers.

Computer-aided estimating systems allow early pricing to start, using the rates contained in the main library of resource costs. When quotations arrive, the resource costs can be updated in the job library. Estimators can make good progress using this approach but must be careful to check that all the prices are confirmed by suppliers (preferably in writing) before the tender is submitted.

Components of a rate

Unit rates are usually a combination of rates for labour, plant, materials and sub-contractors. *Only the direct site cost is included* because management will develop a better understanding of the pricing level if on-costs are dealt with separately. There is a more extreme view that rates should ignore some or all of the following:

1. General site plant such as cranes and plant for materials distribution such as tractors and trailers, dumpers and forklift trucks.
2. Small plant, tools and safety equipment.
3. General labourers assisting craftsmen, unloading materials, distributing materials and driving mechanical plant.
4. Difficult working conditions such as access, restricted space and exposure to the weather.

117

CB CONSTRUCTION LIMITED

Project	Fast Transport Ltd		Trade	Ironmongery
Ref.no	T354	Date 28.6.94	Page	1

Item	Item	Item	Description	Hrs	Total Quants	Unit	LABOUR £8.00/h rate	£	PLANT rate	£	MATERIALS basic	sund	waste	rate	£	LOSC rate	£
3/26A	4/15F		Overhead door closer	1.50	12	nr	12.00	144.00	–	–			2.50			15.00	180.00
3/26B	4/15G		200mm flush bolt	0.75	5	nr	6.00	30.00	–	–			2.50			7.00	35.00
3/26C	4/15F		Mortice dead lock	0.75	8	nr	6.00	48.00	–	–			2.50			9.00	72.00

Fig 11.1 *Estimator's build-up sheet for fixing ironmongery*

The estimator must think about the way in which each operation will be carried out. The following factors must be considered in calculating the cost:

1. Quantity of work to be done.
2. Quality of work and type of finish specified.
3. The degree of repetition.

Many clients assume that unit rates are for all the obligations and risks associated with the work, and in some cases include a statement such as 'the rates inserted in the bill of quantities are to be fully inclusive of all costs and expenses together with all risks, liabilities, obligations given in the tender documents'. Does this mean that a proportion of general expenses should be included in all the rates or can the contractor insert rates for all general obligations and overheads in the preliminaries bill?

Method of measurement

The classification tables in SMM7 set out the work which is to be included in the unit rate. For example, when working space is measured to excavations, the contractor is to allow for additional earthwork support, disposal, backfilling, work below ground level and breaking out. Clearly the estimator must be aware of the coverage rules before pricing the work. With CESMM3, items for excavation include working space as well as upholding sides of excavation and removal of dead services. In addition, bills of quantities often have a preamble (civil engineering work) or rules for measurement (building) which list the changes to the standard measurement rules. A typical example is the statement 'the contractor shall allow all methods necessary to withhold the sides of excavations including where necessary trench sheeting or sheet piling'. This is a significant change to SMM7 because sheet piling is normally measurable under D32 Steel Piling.

Pricing notes

There are many ways to present pricing notes. Standard forms help the estimator to produce clear information which can be read by others. The form shown in Fig. 11.1 would allow an estimator to price labour and plant himself and add rates received from a labour-only sub-contractor when they arrive. A direct comparison can then be made. This is similar to the example given in Fig. 12.1 in the following chapter for the comparison of sub-contractors' rates. The form for the pricing notes in this chapter is typically used for detailed build-ups. Its use has declined with the growth of computing.

Pricing notes are not always clearly presented by estimators. Where time is short, they sometimes produce their notes in the bill of quantities either in the margin or on the facing page. At the adjudication stage, management must examine the bill of quantities because summaries for labour, materials and plant will not be available.

Construction staff need to be aware, however, that any tender notes may be useful to understand the logic used at tender stage but the costs may have been changed by management at the adjudication meeting. A computer system, on the other hand, will produce an up-to-date report of resources with all changes made after the review stages. There is no doubt that the computer reports are quick to produce and can provide comprehensive site budgets and valuations. Very few give reports on the logic used to build up rates, which means that some manual notes or method statements may still be necessary.

Model rate and pricing examples

The way in which unit rates are built up differs from company to company and between trades. Calculations for earthworks, for example, are based on the use of plant, and formwork pricing depends on the making and re-use of shutters. A checklist could be used by trainee estimators or anyone pricing an item for the first time. The 'model rate' calculation given in Fig. 11.2 has more components than any one item would need.

The pricing information sheets given in this chapter contain typical outputs and pricing notes for the categories of work found most often in building and small civil engineering projects.

Most of the data have been expressed in terms of decimal constants which are used for entering resources in a computer application. Unfortunately, this approach gives some strange results and unfamiliar figures. With excavation of trenches, for example, estimators think in terms of how many cubic metres could be dug in one hour, and not the reverse, an output such as $0.20\,\text{hour/m}^3$.

For clarification the following points should also be kept in mind:

1. Most of the examples are for work measured using the rules of SMM7.
2. Each construction organization should decide how to deal with labour and plant in off-loading lorries and distributing materials on site; in either the unit rates or preliminaries.
3. The pricing notes do not bring out the concept of gang sizes. The composition

of a brickwork gang may be two bricklayers assisted by one labourer, in other words, a 2:1 gang. This is written as the time for a bricklayer and half the amount of time for a labourer.

4. The headings SMALL, MEDIUM and LARGE refer to the quantity of an operation.
5. The labour rates used are £8.00 for skilled and £6.50 for unskilled.
6. The outputs for labour and plant represent average times.

Fig 11.2 *Model rate calculations*

Project		Trade	MODEL RATE	Date	
Ref. No.			Unit rate pricing	Sheet No.	

Typical bill description	Z20 Section of construction material Section of construction material fixing to brickwork 160m

item details					analysis				nett
ref:	description	quant	unit	rate	lab	plt	mat	s/c	unit rate
Mat	Unit price from supplier	1	m	2.42			2.42		
	Delivery and packing charges	1/160	item	20.00			0.13		
	Overlap (usually sheet materials)								
	Penetration (usually aggregates)								
	Nails, plugs, screws, adhesives etc	3	nr	0.12			0.36		
	Mortar (usually bricks, blocks & kerbs)								
	Waste - cutting from larger pieces								
	Waste - breakages before fixing								
	Waste - during fixing	0.05	m	2.91			0.15		
	Waste - residue from large packs								
Lab	Unload, store and distribute	0.01	hr	6.50	0.07				
	Craftsman at all-in rate	0.2	hr	8.00	1.60				
	Labourer assistance at all-in rate	0.04	hr	6.50	0.26				
Plt	Electric drill and masonry bit	0.2	hr	0.40		0.08			
	(Small tools and equipment usually								
	priced in preliminaries as a small								
	percentage added to all labour costs)								
	Total net rate				1.93	0.08	3.06		5.07
ADD	a proportion of overheads & profit								
	(the rest will be shown as items								
	in the preliminaries bill)	10	%		0.19	0.01	0.31		0.51
	TOTAL UNIT RATE	1	m		2.12	0.09	3.37		5.58

PRICING INFORMATION		GROUNDWORKS EXCAVATION

SMM7 NOTES	CESMM3 NOTES
Work section D20 **Excavating and filling**	**CLASS E** **EARTHWORKS**

1 Information given in tender documents:
   a. ground water level
   b. details of trial holes or boreholes
   c. live services and features retained	

2 Working space measured separately	Excavation deemed to include working space
3 Excavating below water measured separately	
4 Earthwork support is measured whether needed or not, except to faces ne 0.25m high and faces next to existing structures	Excavation includes upholding sides
5 Interlocking steel piling must be measured (D32)	Piling for temporary works not measured
6 Excavating foundations around piles identified	Excavating foundations around piles identified
7 Underpinning measured in Section D50	Class E includes excavation for underpinning

EXCAVATION		hand dig	small		medium		large	
			JCB3CX	JCB JS150	JCB3CX	JCB JS150	JCB JS150	CAT225
Topsoil		2 - 3	0.30	0.20	0.20	0.15	0.10	0.08
Reduce levels & basements	ne 0.25m	2 - 3	0.30	0.12	0.20	0.11	0.09	0.05
	ne 1.00m	2 - 3	0.20	0.10	0.15	0.09	0.07	0.04
	ne 2.00m	3 - 4	0.20	0.10	0.15	0.08	0.06	0.04
	ne 4.00m	4 - 5	0.25	0.12	0.20	0.11	0.09	0.05
	ne 6.00m		0.30*	0.15	0.25*	0.13	0.11	0.07
Trenches/Pits	ne 0.25m	2 - 3	0.35	0.22	0.25	0.17	0.11	0.07
	ne 1.00m	3 - 4	0.25	0.20	0.20	0.14	0.10	0.06
	ne 2.00m	4 - 5	0.25	0.20	0.20	0.12	0.09	0.05
	ne 4.00m	5 - 7	0.30	0.25	0.25	0.20	0.12	0.07
	ne 6.00m		0.35*	0.30	0.30*	0.25	0.15	0.09

Average outputs - hr/m³

* May be beyond range of machine

BREAKING OUT EXISTING MATERIALS hr/m³	ROCK	CONC	R CONC	MASONRY	SURFAC'G	these outputs are for breaking only ADD the following:
Compressor & labourers	3.00	2.00	3.50	1.50	0.75	25% for trench work
JCB 3CX and breaker	0.50	0.40	0.55	0.25	0.20	25% to excavation rate
JCB 812 & breaker	0.35	0.25	0.40	0.15	0.10	25% to loading rate
CAT 225 and breaker	0.30	0.20	0.30	0.10	0.08	25% to removal rate

Excavation outputs are normally expressed as m³/h. These tables use decimal constants for computer applications

Excavation outputs depend on:	
Quantities	small, medium and large in the table is a guide to quantity of excavation
Ground conditions	the data above are based on 'normal' ground conditions (firm clay)

Bucket size	outputs based on: JCB3CX (backhoe/loader) with a bucket capacity of 0.30m³
	   JCB JS150 (backacter) with a bucket capacity of 0.60m³
	   CAT225 (backacter) with a bucket capacity of 1.20m³

Location	outputs assume reasonable access for plant and lorries
Disposal	where lorries have clear access, the above outputs are sufficient to excavate and load
Trimming	the outputs provide for trimming if labour is included in the excavation rate; trimming should be priced separately for large areas and sloping surfaces

PRICING INFORMATION		GROUNDWORKS DISPOSAL AND FILLING

SMM7 NOTES	CESMM3 NOTES
Work section D20 **Excavation and filling**	**CLASS E** **EARTHWORKS**
1 Disposal off site is stated	Disposal of excavated material is deemed to be
2 Disposal on site is stated	off site unless otherwise stated
3 Only design-imposed locations are stated	The location for material for disposal on site is given
4 Only design-imposed handling provisions are stated	Double handling measured where expressly required
5 Kind and quality of fill materials are stated	Materials for imported filling are given
6 Compaction of filling is measured separately	Filling is deemed to include compaction
7 The filling quantity is the volume of void filled	
8 Filling measured m³ and compaction m²	Filling to stated thicknesses measured m²
9 Filling thickness is that after compaction	
	Penetration of filling over 75mm deep is measured

LOADING OF EXCAVATED MATERIAL

outputs for loading lorries/dumpers

	m³/h
JCB3CX	10
JCB JS150	15
CAT 225	20 +

REMOVAL OF EXCAVATED MATERIAL

	tip located	
	on site	off site
Average speed to tip	10mph	15mph
Average time on tip	3min	6min
Average speed to return	15mph	20mph

DEPOSITION AND COMPACTION OF FILLING MATERIALS		Output hr/m³					
		JCB3CX + 2 Labs & roller		JCB JS150 + 2 Labs & roller		CAT 943 + 2 Labs & roller	
quantities		*small*	*medium*	*small*	*medium*	*medium*	*large*
Filling to excavations	ne 0.25m	0.25	0.17	0.12	0.10	0.08	0.05
	over 0.25m	0.17	0.12	0.10	0.08	0.07	0.04
Making up levels	ne 0.25m	0.25	0.17	0.12	0.07	0.05	0.03
	over 0.25m	0.17	0.12	0.10	0.05	0.04	0.03
Blinding surfaces		0.33	0.20	0.25	0.20		
		Output hr/m³					
		1 Lab & roller/rammer		1 lab & tandem roller		CAT 943 & towed roller	
Compacting open excavation/ground		0.10	0.05	0.05	0.03	0.02	0.01
Compacting filling [if not priced above]		0.20	0.10	0.06	0.04	0.02	0.01
Compacting under foundations		0.20	0.10				

Outputs are normally expressed as m³/h. This table uses decimal constants for computer applications

Material from site spoil heaps will need to be loaded and transported to the filling site

Conversion factors	Ashes	1.30	Gabion stone	1.50
including	Blast furnace slag	2.10	Crushed limestone	1.95
consolidation	Sand	1.75	Scalpings	2.10
t/m³	Stone dust	1.75	DOT type 1	2.30

			£	p
	D20 EXCAVATING AND FILLING			
	Excavating			
	Topsoil for preservation			
A	275 average depth	380 m²		
	To reduce levels			
B	1m maximum depth, commencing 275 below existing ground level	246 m³		
	Trenches exceeding 300 wide			
C	1m maximum depth, commencing 600 below existing ground level	38 m³		
	Extra over excavation irrespective of depth for breaking out			
D	rock (approximate)	26 m³		
	Working space allowance to excavations			
E	trenches, backfilling with selected excavated material	94 m²		
	Earthwork support			
	To faces of excavation			
F	2m maximum depth, distance between opposing faces not exceeding 2m	140 m²		
	Disposal			
	Excavated material			
G	off site	237 m³		
	Selected excavated material			
	Filling to excavations			
H	over 250 thick	47 m³		
	Hardcore as D20.M010			
	Filling to make up levels			
I	over 250 thick, obtained off site	252 m³		

To collection

CB CONSTRUCTION LIMITED   PRICING NOTES

Project		Trade	EXCAVATION	Date	
Ref. No.			Unit rate pricing	Sheet No.	1

item details					analysis				nett
ref:	description	quant	unit	rate	lab	plt	mat	s/c	unit rate
	The type of plant to be used should be selected by examining the								
	nature of the ground and quantities for excavation and disposal								
	Page 3/1 of the bill of quantities would be considered in relation to								
	other excavation in the works such as drainage and external works								
	The total excavation on this page is :								
	topsoil 380 x 0.275 =	105	m³						
	to reduce levels	246	m³						
	trenches	38	m³						
		389	m³						
	The total disposal on this page is :								
	disposal off site	237	m³						
	filling to excavations	47	m³						
	topsoil retained on site	105	m³						
		389	m³						
	For a machine excavating at 10m³/h (on average) there would appear								
	to be at least a week of work . The additional costs of transporting								
	a larger (backacter) machine is justified because it will be needed								
	to break out rock and place filling materials								
A	Topsoil 275mm deep								
	JCB JS150	0.15	hr	17.00		2.55			
	Banksman	0.15	hr	6.50	0.98				
	Consider lorry or dumper if spoil								
	to be taken away from building area								
		1	m³		0.98	2.55			3.53
	Topsoil 275mm deep (x 0.275)	1	m²		0.27	0.70			0.97
B	Excavating to reduce levels								
	JCB JS150	0.09	hr	17.00		1.53			
	Banksman	0.09	hr	6.50	0.59				
	Excavating to reduce levels	1	m³		0.59	1.53			2.12

CB CONSTRUCTION LIMITED   PRICING NOTES

Project		Trade	EXCAVATION	Date	
Ref. No.			Unit rate pricing	Sheet No.	2

item details					analysis				nett
ref:	description	quant	unit	rate	lab	plt	mat	s/c	unit rate
C	Excavating trenches								
	JCB 812	0.14	hr	17.00		2.38			
	Banksman	0.14	hr	6.50	0.91				
	Labourer trimming	0.14	hr	6.50	0.91				
	Excavating trenches	1	m³		1.82	2.38			4.20
D	EXTRA breaking out rock								
	Assuming 20% in red lev and 80% in trenches								
	JCB 812 and breaker [20% .35 hr]	0.07	hr	25.00		1.75			
	JCB 812 and breaker [80% .44 hr]	0.35	hr	25.00		8.75			
	Add 25% to excavation rates red lev				0.07	0.19			
	Add 25% to excavation rates trench				0.23	0.30			
	Add 25% to disposal rate	0.25	m³	8.00		2.00			
	EXTRA for breaking out rock	1	m³		0.30	12.99			13.29
E	Working space allowance								
	Excavation as for tenches	1	m³		1.82	2.38			
	Assume 75% filling and 25% disposal								
	JCB 812 and roller [75% of 0.08hr]	0.06	hr	19.00		1.14			
	Labourers (2nr) [75% of 0.16hr]	0.12	hr	6.50	0.78				
	Additional earthwork support	not priced							
	Additional disposal	0.25	m³	8.00		2.00			
		1	m³		2.60	5.52			8.12
	Assuming average thickness is 250mm								
	Working space allowance	1	m²		0.65	1.38			2.03
F	Earthwork support								
	For shallow trenches support may not be required (nil rate)								
	but the trenches may have sloping sides								
	Once an assessment is made of the average over-excavation								
	the working space rate (above) can be used								
	Assuming average thickness is 300mm								
	Earthwork support	1	m²		0.78	1.66			2.44

working space — volume if no earthwork support

CB CONSTRUCTION LIMITED   PRICING NOTES

Project		Trade	EXCAVATION	Date	
Ref. No.			Unit rate pricing	Sheet No.	3

ref:	description	quant	unit	rate	lab	plt	mat	s/c	nett unit rate
	item details					**analysis**			**nett**
G	**Disposal off site**								
	In this case it is assumed that 75% of material								
	can be loaded directly into lorries at the time of excavation								
	This means that 25% is loaded as a separate operation								
	JCB JS150 [15m³/hr x 25%]	0.017	hr	17.00		0.29			
	The speed of loading is less for material loaded								
	directly at the time of excavation, say 10m³/hr								
	The average rate of loading is therefore:								
	25% at 15m³/hr and 75% at 10m³/hr = 11.25m³/hr								
	The other assumptions made for the calculation are:								
	Lorry capacity 16T	6.4m³							
	Distance to tip	5m							
	Tip charges per load	£15.00							
	Round trip calculation								
	Load 6.4m³ at 11.25m³/hr	34	min						
	Haul to tip at 15m/hr	20	min						
	Time on tip	6	min						
	Return to site at 20m/hr	15	min						
	Total	75	min						
	So each lorry will achieve 60/75 = 0.80 trips per hr								
	and carry 6.4 x 0.80 = 5.12 m³/hr								
	If the maximum speed of loading is 15m³/hr,								
	three lorries are needed at £18.00 per hour								
	Lorry cost is therefore 3 x 18.00 = £54.00/hr								
	The average rate of disposal is 11.25m³/hr								
	Lorries	0.09	hr	54.00		4.86			
	Tip charges £15.00 + 6.4m³					2.35			
	Disposal off site	1	m³			7.50			7.50

CB CONSTRUCTION LIMITED   PRICING NOTES

Project		Trade	EXCAVATION	Date	
Ref. No.			Unit rate pricing	Sheet No.	4

item details					analysis				nett
ref:	description	quant	unit	rate	lab	plt	mat	s/c	unit rate
H	Filling with selected excavated material								
	Assuming transport over short distances can								
	be provided by a site dumper, and an allowance								
	has been made for this in the preliminaries:								
	JCB JS150 and roller	0.08	hr	19.00		1.52			
	Labourers (2nr)	0.16	hr	6.50	1.04				
	Filling	1	m³		1.04	1.52			2.56
I	Hardcore filling								
	Hardcore price from supplier	1.95	t	5.80			11.31		
	Waste	0.19	t	5.8			1.10		
	(Penetration into the ground would be								
	considered for hardcore beds under 250mm)								
	JCB JS150 and roller	0.05	hr	19.00		0.95			
	Labourers (2nr)	0.10	hr	6.50	0.65				
	Hardcore filling	1	m³		0.65	0.95	12.41		14.01

PRICING INFORMATION		IN SITU CONCRETE

SMM7 NOTES	CESMM3 NOTES
Work section E10 **In situ concrete**	**CLASS F** **IN SITU CONCRETE**
1 Kind and quality of materials and mixes stated	Concrete mix may be related to BS 5328, or a mix
2 Tests of materials and finished work stated	designed by the contractor, or a mix prescribed
3 Limitations on pouring methods stated	in the specification; with items given separately
4 Methods of compaction and curing stated	for provision and placing of concrete
5 Requirements for beds laid in bays to be given	
6 Concrete assumed to be as struck or tamped finish	Finishes to concrete measured separately
7 Concrete measured net with no deduction for :	Volume of concrete includes that occupied by:
..reinforcement, sections under 0.50m², voids under	reinforcement, cast-in items ne 0.1m³, rebates,
.. 0.05m³	grooves and chamfers ne 0.01m², large and small
8 Details of concrete sections given on drawings	voids, and joints in in-situ concrete
9 Beds include blinding, plinths and thickenings	Placing concrete in blinding measured separately

	waste %	Output - operative hrs/m³					
		small		medium		large	
		plain	reinf	plain	reinf	plain	reinf
Mass filling	10.0	1.45		1.25		1.00	
Foundations	7.5	1.65	2.00	1.40	1.65	1.10	1.30
Ground beams	5.0	2.40	2.90	2.00	2.40	1.60	1.90
Isolated foundations	7.5	1.65	2.00	1.50	1.75	1.25	1.50
Blinding beds	35.0	2.75		2.40		2.00	
Beds ne 150mm	10.0	1.35	1.65	1.30	1.55	1.20	1.45
150 - 450mm	7.5	1.20	1.45	1.15	1.35	1.05	1.25
over 450mm	5.0	1.10	1.30	1.00	1.20	0.90	1.10
Slabs ne 150mm	5.0		4.00		3.00		2.50
150 - 450mm	5.0		3.00		2.50		2.25
over 450mm	5.0		2.50		2.00		1.50
Troughed slabs	5.0		3.50		3.00		2.50
Walls ne 150mm	7.5	3.50	4.20	2.75	3.30	2.00	2.40
150 - 450mm	5.0	2.80	3.35	2.30	2.75	1.80	2.15
over 450mm	5.0	2.00	2.40	1.80	2.15	1.60	1.90
Filling hollow walls 50mm th	20.0	7.00	8.00	6.00	7.00	5.00	6.00
Filling hollow walls 75mm th	15.0	6.00	7.00	5.00	6.00	4.00	5.00
Beams	7.5	4.50	5.40	3.50	4.20	2.50	3.00
Beam casings	10.0	4.95	6.00	3.85	4.65	2.75	3.30
Columns	7.5	4.50	5.40	3.50	4.20	2.50	3.00
Column casings	10.0	4.95	6.00	3.85	4.65	2.75	3.30
Staircases	7.5	3.60	4.30	3.00	3.60	2.40	2.90
Upstands and kerbs	10.0	6.00	7.00	5.00	6.00	4.00	5.00

SLOPING ITEMS	Add 5% to labour for concrete laid up to 15°, and 10% for concrete over 15°
LABOUR RATE	The effective rate for an operative is found by dividing the cost of a concrete gang by the number of
	operatives in the gang.
WASTE	Waste includes losses due to small quantities and irregular levels for beds and blinding
	Part load charges should be considered for very small quantities
CURING	The outputs include labour for protecting and curing fresh concrete
REINFORCEMENT	Consider concrete saving for members reinforced over 5% by volume

CB CONSTRUCTION LIMITED   PRICING NOTES

Project		Trade	IN SITU CONCRETE	Date	
Ref. No.			Unit rate pricing	Sheet No.	

Typical bill description	E10 IN SITU CONCRETE Reinforced in situ concrete; mix B, 20mm aggregate; beds thickness 150-450mm 160m³

item details					analysis				nett
ref:	description	quant	unit	rate	lab	plt	mat	s/c	unit rate
	Hourly rate for concrete gang:								
	Working ganger	1	hr	7.00	7.00				
	Labourers (4nr)	4	hr	6.50	26.00				
	Carpenter in attendance	1	hr	7.00	7.00				
	* Poker vibrator (2nr)	2	hr	1.80		3.60			
	* Concrete pump	1	hr	38.00		38.00			
	Rate for concrete gang	1	hr		40.00	41.60			81.60
	Effective rate for one operative (+5)	1	hr		8.00	8.32			16.32
Mat	Concrete price from supplier	1	m³	39.90			39.90		
	Waste	0.075	m³	39.90			2.99		
Lab	Concrete operative	1.25	hr	8.00	10.00				
Plt	* Vibrator and pump	1.25	hr	8.32		10.40			
	Rate for in situ concrete	1	m³		10.00	10.40	42.89		63.29
	* (plant may be priced in prelims)								

CB CONSTRUCTION LIMITED   PRICING NOTES

Project		Trade	IN SITU CONCRETE	Date	
Ref. No.			Operational pricing	Sheet No.	

> **Typical bill description**
>
> E10 IN SITU CONCRETE
>
> Reinforced in situ concrete; mix B, 20mm aggregate; beds thickness 150-450mm 160m³

item details					analysis				nett
ref:	description	quant	unit	rate	lab	plt	mat	s/c	unit rate
	For a concrete slab 40 x 20m								
	Assume 1nr bay 4m wide cast per day								
	Volume cast = 40 x 4m x .2m thick =	32	m³						
Lab	Ganger	9	hr	7.00	63.00				
	Labourers (4nr)	36	hr	6.50	234.00				
	Carpenter	4.5	hr	7.00	31.50				
Plt	* Poker vibrator (2nr)	18	hr	1.80		32.40			
	* Concrete pump	9	hr	38.00		342.00			
Mat	Concrete price from supplier	32	m³	39.90			1276.80		
	Waste	2.4	m³	39.90			95.76		
	Rate for one bay	32	m³		328.50	374.40	1372.56		2075.46
	Rate for in situ concrete (+32)	1	m³		10.27	11.70	42.89		64.86
*	(plant may be priced in prelims)								

| PRICING INFORMATION | | FORMWORK |

SMM7 NOTES	CESMM3 NOTES
Work section E20	**CLASS G**
Formwork for in situ concrete	**CONCRETE ANCILLARIES [Formwork]**
1 Basic finish is given	Finishes described as rough, fair or other
.... where not at the discretion of the contractor	
2 Plain formwork measured separately	Formwork deemed to be plane areas > 1.22m wide
.... where no steps, rebates, pockets etc	
3 Rules distinguish between left in and permanent	Formwork left in shall be so described
4 Formwork is measured where temporary	Formwork is measured where temporary
support to fresh concrete is necessary	support to fresh concrete is necessary
5 Tender documents give sizes and positions of	
members, and loads in relation to casting times	
6 Radii stated for curved formwork	Radii stated for curved formwork
7 Formwork to members of constant cross section	Formwork to members of constant cross section
measured square metres	measured in linear metres

LABOUR OUTPUTS	Output - carpenter hr/m²					
	small		medium		large	
	make	F&S	make	F&S	make	F&S
Foundations ne 250mm	1.10	1.80	0.90	1.60	0.80	1.50
250-500mm	1.00	1.70	0.80	1.50	0.70	1.40
500mm-1.00m	0.90	1.60	0.70	1.40	0.60	1.30
over 1.00m	0.80	1.50	0.60	1.30	0.50	1.20
Edges of beds ne 250mm	1.20	1.90	1.00	1.70	0.90	1.60
250-500mm	1.10	1.80	0.90	1.60	0.80	1.50
Edges of susp slabs ne 250mm	1.30	2.40	1.10	2.20	1.00	2.00
250-500mm	1.20	2.30	1.00	2.10	0.90	1.90
Sides of upstands ne 250mm	1.20	2.20	1.00	2.00	0.90	1.90
250-500mm	1.10	2.10	0.90	1.90	0.80	1.80
Soffits of slabs horizontal		1.20*		1.00*		0.90*
sloping ne 15°		1.30*		1.10*		1.00*
sloping over 15°		1.40*		1.20*		1.10*
Soffits of troughed slabs		1.20*		1.00*		0.90*
Walls	1.20	1.70	1.00	1.50	0.90	1.40
Walls over 3.0m	1.20	1.80	1.00	1.60	0.90	1.50
Beams isolated regular shape	1.30	2.20	1.10	2.00	1.00	1.80
irregular shape	1.40	2.40	1.20	2.20	1.10	2.00
Beams attached regular shape	1.40	2.30	1.20	2.10	1.10	1.90
irregular shape	1.60	2.50	1.40	2.30	1.30	2.10
Columns isolated regular shape	1.00	1.50	0.80	1.30	0.70	1.10
irregular shape	1.20	1.60	1.00	1.40	0.90	1.20
Columns attached regular shape	1.20	1.80	1.00	1.60	0.90	1.40
irregular shape	1.40	1.70	1.20	1.70	1.10	1.50

The carpenter rate should include part of a labourer's time for handling materials

Items marked * need a carpenter rate plus a full labourer's rate to erect falsework

'Make' applies to timber shutters. Reduce for hired equipment or proprietary systems

ADD 0.15 hr/m² for fix and strike to walls with formwork one side

A small reduction for formwork LEFT IN is balanced by the labour costs in making

PRICING INFORMATION		FORMWORK MATERIALS AND EQUIPMENT

Timber formwork		units	Founds	Edges	Soffits	Walls	Beams	Columns
Sheet material	Plywood	m²/m²	1	1	1	1	1	1
Timber		m³/m²	0.04	0.06	0.05	0.05	0.05	0.05
ADD 10% waste to plywood and timber								
Bolts and nails		kg/m²	1.0	1.0		1.0	1.5	1.5
Surface preparation	(consider varnish to plywood)							
Number of uses *			high	medium	med/low	medium	medium	high
Falsework and equipment	(see examples)							
Consumables	Mould oil	l/m²	0.3	0.3	0.3	0.3	0.3	0.3
	Buried fixings	nr/m²				1.0		
	Nails	kg/m²	1.0	1.0	1.5	1.0	1.5	0.5

* Number of uses

1 Examine drawings and programme to determine the degree of repetition

2 Consider the standard of surface finish required

3 Consult programme to find time constraints

4 Investigate better quality shutters to increase uses

5 Will there be a salvage value after the work is finished

Key : high = over 6 uses
medium = 3 to 6 uses
low = under 3 uses

Standard timber shutters can be used up to six times without too much repair and maintenance At tender stage it can be assumed that the saving from additional uses is countered by the additional costs of repairs.

On the other hand, more than six uses can be achieved with higher quality shutter materials or applied protective coatings

Proprietary formwork systems

The following equipment can be hired for concrete work with little repetition, such as a large machine base or a retaining wall which must be cast in one pour.

Steel or ply-faced panels (pans)

Angles

Soldiers, walings and push/pull props

Tie rods and accessories

Radius panels and curved waling tubes

Equipment for foundations

Ground beams and machine bases may need telescopic props

Road forms are a useful alternative for strip footings, blinding and edges of beds

Equipment for soffits

A proprietary falsework system is usually used to support soffit formwork. A weekly rate can be obtained as a rate per m² of soffit depending on height and load to be carried

Standard profile GRP trough moulds can be hired

Non-standard profiles are purchased

The choice is dictated by the design

Expanded polystyrene trough moulds are cheaper than GRP but deteriorate quicker

Equipment for walls

Walings, soldiers, push/pull props and shutter ties

Lifting equipment - beam, chains and shackles

Equipment for columns

For a column 0.40 x 0.40 x 3.30m high

Timber formwork will usually need:

Clamps at ave 500mm centres	2 sets/m	1.2sets/m²
Telescopic props No.3	4nr/col	0.76nr/m²

For a column 400mm dia there are three options:

1. Cardboard tube for single use

2. GRP shutters for multiple uses

3. Steel shutters for hundreds of uses

CB CONSTRUCTION LIMITED   PRICING NOTES

Project		Trade	FORMWORK	Date	
Ref. No.			Foundations	Sheet No.	

> **Typical bill description**
>
> E20 Formwork for in situ concrete
>
> Formwork with basic finish specification type A to sides of foundations; 250 to 500mm high 60m

item details					analysis				nett
ref:	description	quant	unit	rate	lab	plt	mat	s/c	unit rate
	First calculate cost of 1m² of shutter								
Mat	Plywood	1	m²	5.60			5.60		
	Waste 10%	0.1	m³	5.60			0.56		
	Timber	0.04	m³	195.00			7.80		
	Waste 10%	0.004	m³	195.00			0.78		
	Bolts and nails	1	kg	1.00			1.00		
	Surface preparation	nil							
Lab	Carpenter	0.8	hr	8.00	6.40				
	Labourer (one for four carpenters)	0.25	hr	6.50	1.63				
	Shutter cost	1	m²		8.03		15.74		23.77
Mat	Cost per use assuming 6 uses	÷ 6			1.34		2.62		3.96
	Equipment (say 4nr props/m²)	4	nr	0.20		0.80			
	Consumables - mould oil	0.3	l	0.80			0.24		
	Consumables - nails	1	kg	1.00			1.00		
	Consumables - buried fixings	nil							
Lab	Labour fix and strike								
	Carpenter	1.5	hr	8.00	12.00				
	Labourer (one for four carpenters)	0.375	hr	6.50	2.44				
	Formwork rate for one m²	1	m²		15.78	0.80	3.86		20.44
	The average height of formwork will be found from an examination of the drawings; on the other hand								
	a shutter may be made to suit the maximum height which in this case is 500mm								
	Rate for 500mm high shutter	1	m		7.89	0.40	1.93		10.22

CB CONSTRUCTION LIMITED

PRICING NOTES

Project		Trade	FORMWORK	Date	
Ref. No.			Soffits of troughed floors	Sheet No.	

> **Typical bill description**
>
> E20 Formwork for in situ concrete
>
> Formwork with basic finish type A to soffits of troughed slabs; profile as detail 1 on drawing D338 550mm thick; ribs at 900mm crs; 3.0 to 4.5m high to soffit.... 660m²

ref:	description	quant	unit	rate	lab	plt	mat	s/c	nett unit rate
	item details				analysis				nett
*	Price from supplier for expanded polystyrene core moulds								
	is £35.00/m delivered to site with a maximum of 4 uses								
Mat	Rate for moulds at 900 crs	1.11	m	35.00			38.85		
	Plywood for continuous deck	1	m²	5.60			5.60		
	Waste 5%	0.05	m²	5.60			0.28		
	Timber packing and sole plates	0.04	m³	195.00			7.80		
	Waste 5%	0.002	m³	195.00			0.39		
	Nails	nil							
	Surface preparation	nil							
							52.92		
Mat	Cost per use assuming 4 uses						13.23		
Plt	Falsework from specialist	6	wks	1.20		7.20			
Lab	Labourer erect falsework	0.7	hr	6.50	4.55				
	Carpenter f&s deck and troughs	1	hr	8.00	8.00				
	Labourer (one for four carpenters)	0.25	hr	6.50	1.63				
	Labourer dismantle falsework	0.3	hr	6.50	1.95				
	Rate for troughed formwork	1	m²		16.13	7.20	13.23		36.56
*	This price was calculated by adding the costs of the moulds								
	needed to make a typical pour, including end pieces.								
	The contractor's programme will show that sufficient moulds								
	will be needed to prepare the next bay while the first pour is curing,								
	to allow continuity of work								

CB CONSTRUCTION LIMITED   PRICING NOTES

Project		Trade	FORMWORK	Date	
Ref. No.			Columns (operational pricing)	Sheet No.	

Typical bill description

E20 Formwork for in situ concrete

Formwork with smooth finish type B to isolated
columns; circular 340mm diameter; 3 to 4.50m high
to soffit; in 74 nr 274m²

ref:	description	quant	unit	rate	lab	plt	mat	s/c	unit rate
	Quotation from supplier for circular column shutters:								
	£660 per shutter 340mm dia 3.50m long including delivery to site								
	£95 per kicker shutter 150mm high								
Mat	The programme requires three column shutters								
	Column shutters	3	nr	660.00			1980		
	The programme calls for six column kicker shutters								
	Kickers shutters	6	nr	95.00			570		
	Equipment - props (4 x 3 cols)	15	wk	2.40		36			
	Consumables - mould oil	274	m²	0.50			137		
Lab	Carpenter fixes 3 columns per day	25	days	68.00	1700				
	Labourer (one for two carpenters)	12.5	days	55.25	691				
	No credit value taken								
	Total for 74 columns				2391	36	2687		5114
	Rate for circular columns	÷ 274	1	m²	8.72	0.13	9.81		18.66

PRICING INFORMATION	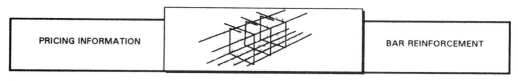	BAR REINFORCEMENT

SMM7 NOTES	CESMM3 NOTES
Work section E30	**CLASS G**
Reinforcement for in-situ concrete	**CONCRETE ANCILLARIES [Reinforcement]**
1 Nominal size of bars is stated	Nominal size of bars is stated
2 Bars classified as straight, bent or curved	Bar shapes not given
3 Links are measured separately	Links not separately identified
4 Lengths over 12m to be given in 3m stages	Lengths over 12m to be given in 3m stages
5 Bar weights exclude rolling margin	Mass of steel assumed to be 7.85t/m³
6 Bar weights inc tying wire, spacers & chairs	Reinf items deemed to include supports to bars
.... only when at the discretion of the contractor	
7 Spacers & chairs measured in tonnes	Mass of reinforcement to include mass of chairs
.... where not at the discretion of the contractor	
8 Location of bars not given in description	Location of bars not given in description
9 Details of conc members given on drawings	

Bar size mm	MASS kg/m	WASTE %	WIRE kg/t	SPACERS nr/t	UNLOAD BARS hrs/t	FIXING TIMES (total hrs/t)			
						FOUNDS	SLABS & BEAMS	LINKS	SITE CUT AND BEND
6	0.222	4.0	14	60	3.7	40	42	56	28
8	0.395	3.5	13	55	3.4	33	35	48	23
10	0.616	3.0	11	50	3.2	28	29	40	20
12	0.888	2.5	9	45	3.0	24	25	36	17
16	1.579	2.5	8	40	2.8	22	24	32	15
20	2.466	2.5	7	35	2.5	20	22		14
25	3.854	2.5	5	30	2.3	18	21		13
32	6.313	2.0	4	25	2.2	16	20		11
40	9.864	2.0	3	20	2.0	14	19		10
50	15.413	2.0	3	15	1.8	14	18		10

STRAIGHT BARS	Suppliers will quote lower prices for straight bars
	The fixing times for straight bars can be reduced by approximately 10-15%
DELIVERY COSTS	Basic prices normally include delivery costs
	Small loads can attract a delivery charge typically £25 for loads under 8 tonnes
SPECIAL SHAPES	Preferred shapes to BS 4466 normally included in bar prices
LONG LENGTHS	Additional charges are made for lengths over 12m
SITE CUTTING	Bars are rarely cut and bent on site except in the case of late design information
	An additional cutting waste would be needed for site cut bars
NETT WEIGHTS	Will steel be charged at calculated weight or weight delivered?

CB CONSTRUCTION LIMITED

PRICING NOTES

Project		Trade	BAR REINFORCEMENT	Date	
Ref. No.				Sheet No.	

Typical bill description

```
E30 REINFORCEMENT FOR IN SITU CONCRETE

Reinforcement bars grade 460 to BS 4449
Bars, 16mm nominal size, bent.....7.82 t
```

ref:	description	quant	unit	rate	lab	plt	mat	s/c	nett unit rate
	item details				**analysis**				**nett**
Mat	Steel per tonne as quotation	1	t	286.00			286.00		
	Waste	0.025	t	286.00			7.15		
	Wire	8	kg	0.80			6.40		
	Spacers	40	nr	0.20			8.00		
Lab	Unload & distribute (plant in prelims)	2.5	hr	6.50	16.25				
	Skilled fixers	22	hr	8.00	176.00				
	Labourer to assist (non-productive)	5.5	hr	6.50	35.75				
	Rate for bar reinforcement	1	t		228.00		307.55		535.55

PRICING INFORMATION		FABRIC REINFORCEMENT

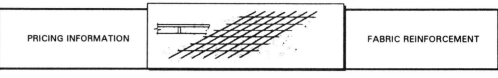

SMM7 NOTES	CESMM3 NOTES
Work section E30 **Reinforcement for in-situ concrete**	**CLASS G** **CONCRETE ANCILLARIES [Reinforcement]**
1 Fabric reference and weight/m² is stated	Fabric ref and weight/m² stated in 2kg/m² bands
2 Laps are not measured	Laps are not measured
3 Laps between sheets are stated	Laps between sheets not stated
4 Fabric inc tying wire, cutting, bending, spacers	Fabric items deemed to include supports
& chairs when at the discretion of the contractor	Supports to top fabric is measured
5 Location of fabric not given in description	Location of fabric not given in description

BS Ref.	MASS kg/m²	UNLOAD FABRIC hr/m²	FIXING TIMES (total hr/m²)					FIXING (hr/m)	
			LARGE BEDS	SMALL BEDS	SLABS	WALLS	BEAMS & COLS	RAKING CUTTING	CIRC CUTTING
A393	6.16	0.02	0.06	0.11	0.14	0.20	0.67	0.40	0.67
A252	3.95	0.01	0.05	0.09	0.12	0.17	0.55	0.32	0.53
A193	3.02	0.01	0.04	0.08	0.10	0.13	0.40	0.25	0.40
A142	2.22	0.01	0.03	0.06	0.08	0.11	0.32	0.22	0.32
A98	1.54	0.01	0.03	0.05	0.07	0.08	0.25	0.17	0.25
B1131	10.90	0.04	0.11	0.17	0.25	0.35	0.90	0.70	1.00
B785	8.14	0.03	0.08	0.14	0.18	0.26	0.80	0.50	0.70
B503	5.93	0.02	0.07	0.11	0.14	0.20	0.67	0.40	0.67
B385	4.53	0.02	0.06	0.10	0.13	0.18	0.60	0.36	0.60
B283	3.73	0.01	0.05	0.08	0.10	0.13	0.40	0.25	0.40
B196	3.05	0.01	0.04	0.06	0.08	0.10	0.30	0.20	0.30
C785	6.72	0.02	0.07	0.12	0.15	0.22	0.80	0.45	0.80
C636	5.55	0.02	0.06	0.11	0.14	0.20	0.70	0.41	0.70
C503	4.34	0.02	0.06	0.10	0.13	0.18	0.60	0.36	0.60
C385	3.41	0.01	0.05	0.08	0.10	0.13	0.45	0.30	0.45
C283	2.61	0.01	0.04	0.06	0.08	0.11	0.32	0.22	0.32
D98	1.54	0.01	0.03	0.05	0.07	0.10	0.30	0.18	0.25
D49	0.77	0.01	0.02	0.04	0.05	0.08	0.20	0.14	0.20

ALLOWANCE FOR WASTE	Large areas		2.50%	For a high proportion of cut sheets
	Small areas		5.00%	the waste must be calculated
ALLOWANCE FOR LAPS	150 laps	10%		
	225 laps	16%		
	300 laps	22%		
	400 laps	31%		
WIRE AND SPACERS	Large areas		2.50%	
	Small areas		5.00%	
CHAIRS	Typically		0.3-0.5 kg/m2	
OPERATIONAL CHECK	Laying fabric reinforcement in some large beds and slabs should be			
	reconciled with labour for laying concrete			

CB CONSTRUCTION LIMITED   PRICING NOTES

Project		Trade	FABRIC REINFORCEMENT	Date	
Ref. No.				Sheet No.	

Typical bill description

E30 REINFORCEMENT FOR IN SITU CONCRETE

Fabric reinforcement
Ref A252; 3.95 kg/m²; 225 minimum laps....556 m²

ref:	description	quant	unit	rate	lab	plt	mat	s/c	nett unit rate
Mat	Price from supplier	1	m²	1.25			1.25		
	Laps	0.16	m²	1.25			0.20		
	Waste	0.025	m²	1.25			0.03		
	Wire and spacers	0.025	m²	1.25			0.03		
	Chairs	0.5	kg	0.28			0.14		
Lab	Unload & distribute	0.01	hr	6.50	0.07				
	Labourer	0.05	hr	6.80	0.34				
	Rate for fabric reinforcement	1	m²		0.41		1.65		2.37
	Fixing gang for fabric in beds may be made up of:								
	4 labourers @ £6.50/hr	26							
	1 skilled fixer @ £8.00/hr	8							
	Effective rate (divide by 5)	34	£	6.80	/hr				

| PRICING INFORMATION | | BRICKWORK |

SMM7 NOTES	CESMM3 NOTES
Work section F10 **Brick/Block walling**	**CLASS U** **BRICKWORK, BLOCKWORK AND MASONRY**
1 Type & nominal size brick stated	Type & nominal size of brick stated
2 Thickness of construction stated	Thickness of construction stated
3 Type of mortar stated	Type of mortar stated
4 Type of bond stated	Type of bond stated
5 Facework and number of sides stated	Surface finish and fair facing stated
6 Type of pointing stated	Type of pointing stated
7 Bonding to existing wall stated	Bonds to existing measured separately
8 Building overhand is stated	Building overhand not stated
9 Deemed to include all rough and fair cutting ... where at discretion of contractor	
10 Deemed to include all mortices and chases	Rebates and chases measured separately
11 Deemed to include raking out joints to form key	
12 Deemed to include returns, ends and angles	
13 Deemed to include centering	
14 Walls include skins of hollow walls	Cavity or composite walls stated

LAYING TIMES for bricks 215 x 102 x 65mm		COMMONS	SEMI-ENG	FACINGS	MORTAR (m³/m²)		
					nett exc waste	no frog 15% waste	frog 15% waste
half brick	hr/m²	1.10	1.15	1.30	0.018	0.021	0.023
half brick fair one side	hr/m²	1.30	1.35	1.50	0.020	0.023	0.026
half brick fair both sides	hr/m²	1.50	1.55	1.70	0.022	0.025	0.029
one brick	hr/m²	2.00	2.10	2.40	0.046	0.053	0.060
one brick fair one side	hr/m²	2.20	2.30	2.60	0.048	0.055	0.062
one brick fair both sides	hr/m²	2.40	2.50	2.80	0.050	0.058	0.065
1½ brick	hr/m²	2.70	2.90	3.20	0.074	0.085	0.096
1½ brick fair one side	hr/m²	2.90	3.10	3.40	0.076	0.087	0.099
1½ brick fair both sides	hr/m²	3.10	3.30	3.60	0.078	0.090	0.101
Isolated piers	bricks/hr	30	25	20			
Projections	bricks/hr	45	40	35			
Arches	bricks/hr	25	20	15			
Specials	bricks/hr	25	20	15			
Brick coping	bricks/hr	40	35	30			
Unload by hand (lab) *	hr/thou	1.00	1.25	1.50			
Distribute (lab) **	hr/thou	1.00	1.25	1.50			

1 The above outputs are for a gang of two bricklayers and one labourer (ratio 1:½)
2 Average waste allowance 5%
3 Waste can be more with facework, but reject bricks can be used elsewhere
4 Bucket handle, weather struck and raked pointing can take longer (add 0.15 hr/m²)
  * Unloading can be omitted since most deliveries include mechanical off-loading
  ** Distribution of materials may be priced in the preliminaries

CB CONSTRUCTION LIMITED   PRICING NOTES

Project		Trade	BRICKWORK	Date	
Ref. No.				Sheet No.	

Typical bill description

F10 BRICK/BLOCK WALLING

Facing brickwork (PC £250 per thousand delivered and off-loaded);cement,lime,sand mortar 1:1:6
Vertical walls one brick thick; English bond; facework both sides with weathered joint as work proceeds;.... 334m²

ref:	description	quant	. unit	rate	lab	plt	mat	s/c	nett unit rate
Mat	For 1m³ of mortar 1:1:6								
	Cement	0.26	t	80.00			20.80		
	Lime	0.13	t	85.00			11.05		
	Sand	1.71	t	8.00			13.68		
Lab	Unloading by general gang	see prelims							
	Mixing by bricklayer's labourer								
Plt	Mixer and dumper	see prelims							
	Rate for 1:1:6 mortar	1	m³				45.53		45.53
Mat	Price from supplier (£250/th)	119	br	0.25			29.75		
	Waste	0.05	m²	29.75			1.49		
	Mortar inc 15% waste	0.058	m³	45.53			2.64		
Lab	Distribute bricks (dumper in prelims)	0.17	hr	6.50	1.11				
	Bricklayers	2.80	hr	8.00	22.40				
	Labourer	1.4	hr	6.50	9.10				
	Extra for weathered joint both sides	0.3	hr	8.00	2.40				
	Rate for facing brickwork	1	m²		35.01		33.88		68.89

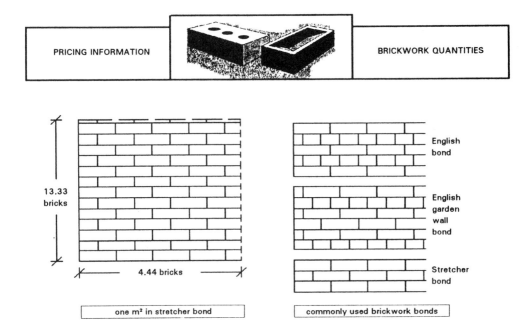

PRICING INFORMATION		BRICKWORK QUANTITIES

13.33 bricks

4.44 bricks

one m² in stretcher bond

English bond

English garden wall bond

Stretcher bond

commonly used brickwork bonds

BRICKS	wall thickness mm	Number of bricks per square metre				
			English bond		English garden wall bond	
		total	facing	commons to rear	facing	commons to rear
half brick	102.5	60				
one brick facework one side	215	119	89	30	74	45
one brick facework both sides	215	119	119		119	
1 ½ brick facework one side	327.5	178	89	89	74	104
1 ½ brick facework both sides	327.5	178	178		148	30

CONSTITUENTS OF MORTAR

SITE MIX CEMENT/LIME/SAND MORTARS BY VOLUME	cement t/m³	lime t/m³	sand t/m³	READY MIX LIME/SAND MORTARS BY VOLUME	cement t/m³	LSM t/m³
Bulk density	1.12-1.60	0.50-0.85	1.35-1.60			
Typical dry density	1.44	0.72	1.60	Typical dry density	1.44	1.85
1:3	0.49		1.62			
1:4	0.39		1.74			
1:5	0.32		1.83			
1:6	0.28		1.88			
1:1:6	0.26	0.13	1.71	1:6	0.24	1.75
1:2:9	0.18	0.18	1.75	1:9	0.17	1.85
1:3:12	0.12	0.18	1.75	1:12	0.13	1.85

PRICING INFORMATION		BRICKWORK SUNDRIES

SMM7 NOTES	CESMM3 NOTES
Work section F30	**CLASS U**
Accessories/Sundry items	**BRICKWORK, BLOCKWORK AND MASONRY**
1 Closing cavities; width of cavities stated	Closing cavities; width of cavities stated
2 Bonding to existing; thickness stated	Bonds to existing work measured 'square'
3 Forming cavities; width and ties stated	Cavity construction stated
4 Damp proof courses measured 'square'	Damp proof courses measured 'linear'
5 Joint reinforcement; width stated	Joint reinforcement; width stated
6 Laps in DPC and joint reinf not measured	Laps in DPC and joint reinf not measured
7 Joints in walls measured where designed	Joints in walls measured where designed
8 Proprietary items	Fixings and ties
9 Pointing flashings incl cutting grooves	

Brickwork labours

Forming cavities	0.03 hr/m²
Closing cavities vert	2.00 hr/m²
Closing cavities horiz	2.00 hr/m²
Bonding to existing	3.50 hr/m²
Prepare wall for raising	1.00 hr/m²
Wedging and pinning	1.25 hr/m²

Designed joints

Fibreboard to joint ne 200	0.15 hr/m
Fibreboard to joint > 200	0.50 hr/m²
One-part mastic 10x10mm	0.20 hr/m
One-part mastic 20x20mm	0.40 hr/m
Two-part sealant 10x10mm	0.30 hr/m
Two-part sealant 20x20mm	0.60 hr/m

Hoist and bed lintels

Small precast concrete	0.15 hr/m
Large precast concrete	0.30 hr/m
Small steel	0.20 hr/m
Large steel	0.25 hr/m

Accessories

Build in butterfly tie	0.01 hr each
Build in twisted tie	0.02 hr each
Build in joist hanger	0.10 hr each
Build in joint reinforcement	0.40 hr/m²

Insulation

Fix 25mm cavity bats (inc clips)	0.20 hr/m²
Fix 50mm cavity bats (inc clips)	0.25 hr/m²
Fix 75mm cavity bats (inc clips)	0.30 hr/m²

Damp-courses

	ne 250	>250
	hr/m	hr/m²
Vertical	0.05	0.30
Raking	0.05	0.30
Horizontal	0.04	0.20
Stepped	0.08	0.40
Laps and waste	15%	10%

Pointing in flashings	0.40 hr/m

PRICING INFORMATION	BLOCKWORK

SMM7 NOTES	CESMM3 NOTES
Work section F10	**CLASS U**
Brick/Block walling	**BRICKWORK, BLOCKWORK AND MASONRY**
1 Type & nominal size of block stated	Type & nominal size/thickness of block stated
2 Thicknes of construction stated	
3 Type of mortar stated	Type of mortar stated
4 Type of bond stated	Type of bond stated
5 Facework and number of sides stated	Surface finish and fair facing stated
6 Type of pointing stated	Type of pointing stated
7 Bonding to existing wall stated	Bonds to existing measured separately
8 Building overhand is stated	
9 Deemed to include all rough and fair cutting	
... where at discretion of contractor	
10 Deemed to include all mortices and chases	Rebates and chases measured separately
11 Deemed to include raking out joints to form key	
12 Deemed to include returns, ends and angles	
13 Deemed to include centering	
14 Walls include skins of hallow walls	Cavity or composite walls stated

	FIXING TIMES for blocks 415 x 215mm		75	100	150	190	200	215
solid	Lightweight blocks	hrs/m²	0.45	0.50	0.60	0.65	0.70	0.80
	Dense concrete blocks	hrs/m²	0.60	0.65	0.80	0.90	1.00	1.15
	Masonry blocks	hrs/m²	0.75	0.80	1.00	1.20	1.25	1.40
hollow	Lightweight blocks	hrs/m²	0.42	0.46	0.55	0.60	0.62	0.65
	Dense concrete blocks	hrs/m²	0.58	0.61	0.67	0.75	0.80	0.85
	Masonry blocks	hrs/m²	0.50	0.58	0.61	0.65	0.67	0.75
	Unload and distibute (lab)	hrs/m²	0.05	0.07	0.10	0.13	0.13	0.14
	Extra for fairface one side	hrs/m²	0.13	0.13	0.13	0.13	0.13	0.13
	Extra for fairface both sides	hrs/m²	0.20	0.20	0.20	0.20	0.20	0.20
	Mortar quants (exc waste)	m³/m²	0.0050	0.0067	0.0100	0.0127	0.0133	0.0144
	Mortar quants (15% waste)	m³/m²	0.0058	0.0077	0.0115	0.0146	0.0153	0.0166

1 The above outputs are for a gang of two bricklayers and one labourer (ratio 1:½)

2 For heavy blocks marked with [   ] allow a labourer with every bricklayer (ratio 1:1)

3 Waste allowance 5% - 7½% (except 10% for fairfaced blockwork and small areas)

4 Add to labour for high walls (30%) dwarf walls (35%) casings (60%) filling openings (70%)

CB CONSTRUCTION LIMITED   PRICING NOTES

Project		Trade	BLOCKWORK	Date	
Ref. No.				Sheet No.	

Typical bill description

F10 BRICK/BLOCK WALLING

Dense concrete blocks 7N;cement and sand mortar 1:3;stretcher bond;vertical walls; facework one side;flush pointing; 150mm thick .... 541 m²

item details					analysis				nett
ref:	description	quant	unit	rate	lab	plt	mat	s/c	unit rate
Mat	For 1m³ of mortar mix 1:3								
	Cement	0.49	t	80.00			39.20		
	Sand	1.62	t	8.00			12.96		
Lab	Unloading by general gang	see prelims							
	Mixing by bricklayer's labourer								
Plt	Mixer and dumper	see prelims							
	Rate for 1:3 mortar	1	m³				52.16		
Mat	Price of BLOCKS from supplier	1	m²	5.50			5.50		
	Waste	0.05	m²	5.50			0.28		
	Mortar inc 15% waste	0.0115	m³	52.16			0.60		
Lab	Unload & distribute	0.1	hr	6.50	0.65				
	Bricklayers	0.8	hr	8.00	6.40				
	Labourer	0.4	hr	6.50	2.60				
	Bricklayer pointing one side	0.13	hr	8.00	1.04				
	Rate for blockwork	1	m²		10.69		6.38		17.07

PRICING INFORMATION		STRUCTURAL TIMBER

SMM7 NOTES	CESMM3 NOTES
Work section G20	**CLASS O & Z**
Carpentry/Timber framing/First fixing	**TIMBER**
1 Kind, quality and treatment of timber stated	Grade or species and treatment stated
2 Sawn or wrot timber stated	Sawn or wrot timber stated
3 All sizes are nominal	Nominal gross cross-sectional areas given
.... unless stated as finished sizes	Thickness of timber stated
4 Method of fixing and jointing given	Method of fixing and jointing not given
.... where not at the discretion of the contractor	
5 Labours on timbers not measured	Boring and cutting not measured
6 Lengths given if over 6m	Length of timber given in one of seven bands Class O

size of member width	size of member depth	sectional area m²	plates floor/roof members	wall or partition members	pitched roof members	gutters fascias eaves soffit	supports butted grounds, battens, firrings, fillets, upstands, drips etc	supports framed
			hr/m	hr/m	hr/m	hr/m	hr/m	hr/m
38	38	0.001	0.09	0.14	0.12	0.18	0.14	0.28
	50	0.002	0.11	0.16	0.14	0.11	0.16	0.36
	75	0.003	0.12	0.20	0.17	0.24	0.20	0.40
	100	0.004	0.13	0.22	0.18	0.26	0.22	0.44
	125	0.005	0.14	0.24	0.19	0.28	0.24	0.48
	150	0.006	0.15	0.26	0.22	0.30	0.26	0.52
	175	0.007	0.16	0.28	0.24	0.32	0.28	0.56
	200	0.008	0.17	0.30	0.25	0.34	0.30	0.60
50	50	0.003	0.12	0.20	0.17	0.24	0.20	0.40
	75	0.004	0.13	0.22	0.18	0.26	0.22	0.44
	100	0.005	0.14	0.24	0.19	0.28	0.24	0.48
	125	0.006	0.15	0.26	0.22	0.30	0.26	0.52
	150	0.008	0.17	0.30	0.25	0.34	0.30	0.60
	175	0.009	0.19	0.34	0.28	0.38	0.34	0.68
	200	0.010	0.21	0.40	0.33	0.42	0.40	0.80
	225	0.011	0.23			0.46		
	250	0.013	0.25			0.50		
	300	0.015	0.30			0.60		
75	100	0.008	0.17	0.29	0.25			
	125	0.009	0.19	0.34	0.28			
	150	0.011	0.23	0.50	0.40			
	175	0.013	0.25	0.55	0.46			
	200	0.015	0.30					
	250	0.019	0.37					
	300	0.023	0.40					
100	100	0.010	0.21	0.40	0.33			
	150	0.015	0.30	0.60	0.50			
	200	0.020	0.37					
	250	0.025	0.42					
	300	0.030	0.45					
150	150	0.023	0.40					
	200	0.030	0.45					
	300	0.045	0.50					

herringbone and block strutting

depth of joists mm	hr/m measured over joists
75	0.22
100	0.25
125	0.26
150	0.27
175	0.28
200	0.30
225	0.32
250	0.35
300	0.40

Nails can be priced in the preliminaries with a sum for sundry fixings. Alternatively, allow 2kg/m³ of timber. For example: if nails cost £1.20/kg the rate for timber 50x100mm would be 2 x 1.2 x .05 x .10 = 1p/m

The above outputs are for each carpenter

Labour assisting carpenters can be added to the all-in rate or priced in the preliminaries

Average waste for structural timbers generally 7.5% .

Outputs will vary significantly depending on the complexity of the work

PRICING INFORMATION		JOINERY

SMM7 NOTES	CESMM3 NOTES
Work section N10 **General fixtures ..** **Work section P20** **Unframed isolated trims ..**	**Class Z** **SIMPLE BUILDING WORKS ..** **Carpentry and joinery**

P20	All timber sizes are nominal sizes unless stated as finished sizes	All timber sizes are nominal sizes unless otherwise stated
	The work is deemed to include ends, angles, mitres,intersections except hardwood items over 0.003m² sectional area	Boring, cutting and jointing not measured
	Kind, quality and treatment of timber stated	Grade or species and treatment of timber stated
	Sawn or wrot timber stated	Sawn or wrot timber stated
	Method of fixing given .. if not at the discretion of the contractor	Method of fixing not given

Fixing softwood skirtings, architraves, trims, window boards etc   Hr/m

size of member		Nailed	Screwed	Plugged &screwed	Screwed &pelleted	size of member		Nailed	Screwed	Plugged &screwed	Screwed &pelleted
19	19	0.10	0.13	0.18	0.23	38	19	0.12	0.15	0.22	0.25
	25	0.10	0.13	0.18	0.23		25	0.12	0.15	0.22	0.25
	32	0.10	0.13	0.18	0.23		32	0.12	0.15	0.22	0.25
	38	0.12	0.15	0.20	0.25		38	0.15	0.19	0.26	0.29
	44	0.12	0.15	0.20	0.25		44	0.15	0.19	0.26	0.29
	50	0.12	0.15	0.20	0.25		50	0.17	0.21	0.28	0.31
	63	0.12	0.15	0.20	0.25		63	0.19	0.24	0.31	0.34
	75	0.15	0.19	0.24	0.29		75	0.23	0.29	0.36	0.39
	100	0.17	0.21	0.26	0.31		100	0.29	0.36	0.43	0.46
	125	0.19	0.24	0.29	0.34		125	0.31	0.39	0.46	0.49
25	19	0.10	0.13	0.19	0.23	50	19	0.12	0.15	0.23	0.25
	25	0.10	0.13	0.19	0.23		25	0.12	0.15	0.23	0.25
	32	0.12	0.15	0.21	0.25		32	0.15	0.19	0.27	0.29
	38	0.12	0.15	0.21	0.25		38	0.17	0.21	0.29	0.31
	44	0.12	0.15	0.21	0.25		44	0.19	0.24	0.32	0.34
	50	0.12	0.15	0.21	0.25		50	0.19	0.24	0.32	0.34
	63	0.15	0.19	0.25	0.29		63	0.24	0.30	0.38	0.40
	75	0.17	0.21	0.27	0.31		75	0.29	0.36	0.44	0.46
	100	0.19	0.24	0.30	0.34		100	0.31	0.39	0.47	0.49
	125	0.23	0.29	0.35	0.39		125	0.4	0.50	0.58	0.60

The above outputs are for each carpenter
ADD 30% to the outputs for fixing hardwood
Average waste allowance is 7.5%. Varies depending on number of short lengths and mitres
ADD for cost of screws

WC CUBICLES	fix
Each door	1.25 hr
Each partition	2.00 hr
Each fascia panel	1.50 hr

KITCHEN UNITS		assemble	fix
Base unit	each	0.65hr	0.50hr
Wall unit	each	0.55hr	0.75hr
Worktop	metre		0.50hr

CB CONSTRUCTION LIMITED   PRICING NOTES

Project		Trade	STRUCTURAL TIMBER	Date	
Ref. No.				Sheet No.	

Typical bill description	G20 Carpentry/Timber framing/First fixing
	Sawn stress graded (SS) timber, pressure impregnated with preservative pitched roof members; 50 x 175mm 174m

item details					analysis				nett
ref:	description	quant	unit	rate	lab	plt	mat	s/c	unit rate
Mat	Timber price from supplier	1	m	1.93			1.93		
	Nails	0.02	kg	1.20			0.02		
	Waste	0.075	m	1.95			0.15		
Lab	Unload and distribute	see prelims							
	Carpenter	0.28	hr	8.00	2.24				
	One labourer assisting four carpenters	0.07	hr	6.50	0.46				
	Unit rate for timber	1	m		2.70		2.10		4.80

PRICING INFORMATION		WINDOWS AND DOORS

SMM7 NOTES	CESMM3 NOTES
Work section L	**CLASS Z**
Windows/Doors/Stairs	**SIMPLE BUILDING WORKS ..**
1 Kind and quality of materials given	Shape, size and limits of work given
2 Details given for treatment, tolerances, jointing, and fixing to vulnerable materials	Items are deemed to include fixings and drilling
3 Bedding and pointing frames measured	
4 Ironmongery, trims, surrounds, glazing and fixings deemed to be included where supplied with the component	Glazing is measured separately. Ironmongery and frames may be included in items for doors and windows where clearly stated
5 Each leaf of multiple doors counted as a door	
6 Approximate weight is given for metal doors	
7 For glass supplied separately see L40	
8 Ironmongery (P21) includes matching screws	Materials stated for ironmongery
9 Nature of base for ironmongery is given	

FIXING TIMBER WINDOWS

		Output - carpenter hr/unit (or perimeter length)					
		casement		box sash		roof windows	
		nr	m	nr	m	nr	m
Windows	ne 2m girth		0.35		0.45		0.50
	2 - 4m girth	1.00		1.40		1.55	
	4 - 6m girth	1.40		1.90		2.20	
	over 6m girth		0.26		0.35		0.40
Bedding frames			0.10				
Pointing frames			0.15				

FIXING TIMBER DOORS AND FRAMES

			Output - carpenter hr/unit including hinges					
			standard		1 hr fire		ledged & braced	
			door	door set	door	door set	door	door set
Doors	small *	m²	1.10	1.70	1.35	2.10	1.00	
	762 x 1981	nr	1.50	2.60	2.00	3.40	1.40	
	838 x 1981	nr	1.60	2.70	2.20	3.60	1.50	
	large	m²	0.90	1.50	1.30	2.00	0.85	

* minimum 0.75hr/door or 1.25hr/door set

			Output - carpenter hr/m		
Frames	(Jambs, heads and sills)		lining & stops	frame & stops	1 hr frame
	38mm thick	m	0.22	0.24	0.26
	50mm thick	m	0.25	0.26	0.30
	63mm thick	m	0.28	0.30	0.35
	75mm thick	m	0.30	0.34	0.40

SPECIFICATION	Specifiers commonly use manufacturers' references to provide technical requirements
LABOUR RATE	The effective rate for a carpenter is found by dividing the cost of a 'carpentry' gang by the number of carpenters in the gang; typically one labourer can service five carpenters
WASTE	Waste mainly due to damage to manufactured components, typically 2½%
	For lengths of doors stops and linings allow 7½%

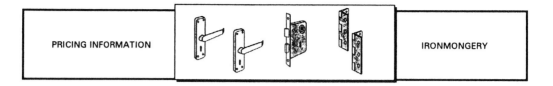

PRICING INFORMATION		IRONMONGERY

* Hinges			Locks and latches	
75mm butts	0.13 hr/pr		Rim latch	0.75 hr each
100mm butts	0.20 hr/pr		Rim dead lock	0.75 hr each
125mm butts	0.25 hr/pr		Mortice latch	0.75 hr each
Rising butts	0.25 hr/pr		Mortice dead lock	0.75 hr each
300mm T hinges	0.30 hr/pr		Mortice deadlock & latch	1.00 hr each
350mm T hinges	0.35 hr/pr		EXTRA for rebated forends	0.50 hr each
Double action spring hinge	1.00 hr each		Cylinder/night latch	0.75 hr each
			Cabinet lock	0.75 hr each
Door closers			Padlock, hasp and staple	0.40 hr each
Perko	1.00 hr each		WC/bathroom indicator bolt	1.00 hr each
Overhead door closer	1.50 hr each			
Single action floor spring	1.50 hr each		Door handles and plates	
Double action floor spring	2.00 hr each		150mm pull handle	0.13 hr each
Door selector stay	0.75 hr each		225mm pull handle	0.17 hr each
			300mm pull handle	0.25 hr each
Bolts			150mm flush handle	0.35 hr each
100mm barrel bolt	0.25 hr each		225mm flush handle	0.60 hr each
150mm barrel bolt	0.33 hr each		300mm flush handle	0.80 hr each
200mm barrel bolt	0.40 hr each		200mm finger plate	0.20 hr each
300mm barrel bolt	0.50 hr each		300mm finger plate	0.25 hr each
100mm flush bolt	0.55 hr each		740 x 225mm kicking plate	0.55 hr each
150mm flush bolt	0.65 hr each		810 x 225mm kicking plate	0.70 hr each
200mm flush bolt	0.75 hr each			
300mm flush bolt	1.00 hr each		Window accessories	
single panic bolt	1.50 hr each		Casement fastener	0.25 hr each
double panic bolt	2.00 hr each		Casement stay	0.25 hr each
			Mortice casement fastener	0.75 hr each
Door accessories			Sash fastener	0.40 hr each
Door security chain	0.20 hr each		Spiral sash balance	0.75 hr each
Door security viewer	0.35 hr each		Sash pulley	0.50 hr each
Lever handles	0.40 hr/pr		Fanlight catch	0.25 hr each
Escutcheon	0.15 hr each			
Letter plate (and slot)	1.25 hr each		Furniture accessories	
100mm cabin hook	0.20 hr each		Cupboard catch	0.25 hr each
Numerals	0.10 hr each		Magnetic catch	0.20 hr each
			Cupboard knob	0.17 hr each
Wall fittings			Cabinet handles	0.20 hr each
Shelf bracket	150mm	0.15 hr each	Curtain track	0.75 hr/m
	200mm	0.17 hr each	Window blind	0.75 hr/m
	250mm	0.20 hr each	Mirror 400 x 600mm	0.60 hr each
Handrail bracket		0.17 hr each		
Toilet roll holder		0.17 hr each	Floor fittings	
Bell push		0.20 hr each	Easyclean socket in concrete	0.40 hr each
Soap dispenser		0.20 hr each	Rubber door stop to timber	0.17 hr each
Hat and coat hooks		0.13 hr each	Rubber door stop to conc	0.30 hr each

WASTE	Allow 2½% for replacement of damaged ironmongery
FIXINGS	Check that ironmongery is supplied with matching screws
	Allow for sundry items such as cavity fixings for hollow backgrounds
* FIXING DOORS	The outputs for fixing doors include the fixing of hinges

CB CONSTRUCTION LIMITED   PRICING NOTES

Project		Trade	DOORS & IRONMONGERY	Date	
Ref. No.				Sheet No.	

Typical bill description

L20 Timber doors/shutters/hatches

Edward Stockley type KL flush doors; plywood faced for painting; 44 x 762 x 1981mm .. 27nr

item details					analysis				nett
ref:	description	quant	unit	rate	lab	plt	mat	s/c	unit rate
Mat	Door price from supplier	1	nr	73.50			73.50		
	Waste	0.025	nr	73.50			1.84		
Lab	Unload and distribute	see prelims							
	Carpenter	1.50	hr	8.00	12.00				
	Labourer assistance	0.20	hr	6.50	1.30				
	Unit rate for door	1	nr		13.30		75.34		88.64

Typical bill description

P21 IRONMONGERY

P & M Winfox Ltd Metric Gold range; single action overhead door closers; Ref: 55006 - to softwood 27nr

Mat	Unit price from supplier inc screws	1	nr	43.25			43.25		
	Waste	0.025	nr	43.25			1.08		
Lab	Unload and distribute	see prelims							
	Carpenter	1.50	hr	8.00	12.00				
	Unit rate for overhead door closer	1	nr		12.00		44.33		56.33

PRICING INFORMATION		PAINTING

SMM7 NOTES	CESMM3 NOTES
Work section M60 **Painting/Clear finishing**	**CLASS V** **PAINTING**
1 Kind and quality of materials stated	Material to be used is stated
2 Nature of base & preparatory work given	Preparation normally deemed to be included
3 Priming, sealing & undercoats enumerated	Number of coats or film thickness given
4 Method of application given	
5 Rubbing down deemed to be included	
6 Work is internal unless otherwise stated	
7 Work in staircase areas and plantrooms stated	
8 Work to ceilings over 3.50m stated (except stairs)	Height of work not given
9 Primary classification is the member painted	Primary classification is the type of paint
10 Secondary classification is the surface features	Secondary classification is the background material
11 No deduction for voids ne 0.50m²	No deduction for voids ne 0.50m²

			Output - operative hr/unit					
			girth > 300mm (m²)		isolated ne 300mm (m)		isolated ne 0.50m² (nr)	
		nr of coats	one	three	one	three	one	three
General surfaces	Emulsion	to plaster	0.09	0.24	0.03	0.08	0.08	0.22
ADD 10% for ceilings		to smooth conc	0.10	0.28	0.03	0.09	0.09	0.25
		to board	0.10	0.26	0.03	0.08	0.09	0.23
		to tex'd paper	0.10	0.26	0.03	0.08	0.09	0.23
		to fair blockwork	0.11	0.30	0.04	0.10	0.10	0.27
	Prepare and prime		0.17		0.02		0.17	
	Undercoat		0.15		0.02		0.15	
	Finishing		0.17		0.03		0.17	
	1 pr, 1 uc, 1 fin			0.45		0.14		0.41
Glazed units	panes	ne 0.10m²	0.24	0.68	0.04	0.22	0.21	0.61
	panes	.10 - 0.50m²	0.20	0.56	0.03	0.18	0.18	0.50
	panes	.50 - 1.00m²	0.18	0.50	0.03	0.16	0.16	0.45
	panes	over 1.00m²	0.16	0.45	0.03	0.14	0.15	0.41
Structural metalwork			0.23	0.65	0.05	0.21	0.21	0.59
Trusses and girders			0.18	0.50	0.06	0.16	0.16	0.45
Radiators			0.19	0.55	0.03	0.17	0.18	0.50
Fencing	Plain open		0.13	0.35	0.02	0.11	0.12	0.32
	Close		0.16	0.45	0.03	0.14	0.15	0.41
	Ornamental		0.23	0.65	0.05	0.21	0.21	0.59
Gutters			0.19	0.55	0.03	0.17	0.18	0.50
Services	(eg. pipes and ducts)		0.23	0.65	0.10	0.21	0.21	0.59

PREPARATION	Washing down, rubbing down, and filling holes included in priming coat
LABOUR RATE	The effective rate for an operative is
HEIGHT ALLOWANCE	ADD 25% for working from ladders
	ADD 15% for working from scaffolding or staging
COVERAGE	Check with paint manufacturer particularly for porous surfaces
	On average 0.07 - 0.08 litres of emulsion or gloss required per m²
	but this could double with surfaces such as blockwork or soft boarding
'OLD' WORK	ADD 25% for painting previously decorated surfaces

CB CONSTRUCTION LIMITED   PRICING NOTES

Project		Trade	PAINTING	Date	
Ref. No.				Sheet No.	

<table>
<tr><td>Typical bill description</td><td colspan="2">M60 PAINTING AND CLEAR FINISHING

Painting concrete - one mist coat and two full coats emulsion paint - as spec M200; general surfaces over 300 girth.. 270m²</td></tr>
</table>

item details					analysis				nett
ref:	description	quant	unit	rate	lab	plt	mat	s/c	unit rate
Mat	Emulsion paint	0.2	l	2.50			0.50		
	Brushes and sundries	1	item	0.10			0.10		
Lab	Painter	0.28	hr	8.00	2.24				
	Unit rate for emulsion painting	1	m²		2.24		0.60		2.84

<table>
<tr><td>Typical bill description</td><td colspan="2">M60 PAINTING AND CLEAR FINISHING

Painting wood - one coat primer, one undercoat and one finishing coat - as spec M300; general surfaces; not exceeding 300 girth.. 64m</td></tr>
</table>

ref:	description	quant	unit	rate	lab	plt	mat	s/c	unit rate
Mat	Paints	0.24	l	3.50			0.84		
	Brushes and sundries	1	item	0.04			0.04		
Lab	Painter - prepare and prime	0.02	hr	8.00	0.16				
	Painter - undercoat	0.02	hr	8.00	0.16				
	Painter - finishing coat	0.03	hr	8.00	0.24				
	Unit rate for oil painting woodwork	1	m		0.56		0.88		1.44

PRICING INFORMATION		DRAINAGE PIPEWORK

SMM7 NOTES	CESMM3 NOTES
Work section R12	**CLASS I**
Drainage below ground	**PIPEWORK - PIPES**
1 Kind and quality and nominal size of pipes stated	Kind and quality and nominal size of pipes stated
2 Method of jointing pipes stated	Method of jointing pipes stated
3 Excavating trenches includes earthwork support,	Excavating trenches includes earthwork support,
.. consolidating trench bottoms, backfilling, and	.. surface preparation, pipework, backfilling, and
.. disposal of surplus excavated materials	.. disposal of surplus excavated materials
4 Backfilling with imported materials stated	Backfilling measured in CLASSES K and L
5 Location for disposal stated	Disposal normally at discretion of contractor
6 Average depth of trench given in 250mm stages	Average depth to invert given in 500mm bands
7 Difficult conditions or locations stated	Pipe locations identified in descriptions
8 Breaking out hard materials & reinstatement given	Crossings and reinstatement given in CLASS K
9 Dimensions of bed and surround given	Hard dig and beds/surrounds given in CLASS L
10 Pipe fittings measured extra over pipework	Pipe fittings and valves measured in CLASS J.

LAYING PIPEWORK		Labour and plant in laying and jointing pipes in trenches hr/m (pipes) hr/nr (fittings)							
		PVC push fit		CLAY push fit or sleeve		CONCRETE push fit		CHANNEL on mortar bed	
		lab	plt	lab	plt	lab	plt	lab	plt
100	pipe	0.10		0.25		0.25		0.40	
	bend	0.14		0.20		0.20		0.50	
	branch	0.28		0.40		0.40		0.55	
150	pipe	0.15		0.30		0.30		0.50	
	bend	0.18		0.25		0.25		0.65	
	branch	0.32		0.45		0.45		0.70	
225	pipe	0.25		0.40		0.40	0.07	0.50	
	bend	0.25		0.35		0.35	0.07	0.90	
	branch	0.39		0.55		0.55	0.10	1.00	
300	pipe	0.35		0.55	0.10	0.55	0.10	0.90	
	bend	0.39		0.55	0.10	0.55	0.10	1.40	
	branch	0.46		0.65	0.15	0.65	0.15	1.55	
375-450	pipe			0.70	0.25	0.70	0.25	1.05	0.20
	bend			0.80	0.25	0.80	0.25	2.10	0.20
	branch			1.05	0.35	1.05	0.35	2.30	0.30
525-600	pipe			0.90	0.35	0.90	0.35	1.45	0.30
	bend			0.90	0.35	0.90	0.35	2.35	0.30
	branch			1.30	0.45	1.30	0.45	2.60	0.40

LABOUR	Outputs are for drainlayer hours per metre of pipe
PLANT	Outputs are for the machine used to excavate the trench
ADDITIONS	ADD to outputs for filled mortar joints (30%); short lengths (30-50%); deep trenches (30-50%)
DEDUCTIONS	DEDUCT for long lengths (20%), shallow trenches (10%)
JOINTS	For joints which are NOT 'push fit', the laying and jointing calculations should be separated.
WASTE	WASTE is normally 5% on pipes and fittings, and 7.5% on short lengths and channels

CB CONSTRUCTION LIMITED   PRICING NOTES

Project		Trade	DRAINAGE EXCAVATION	Date	
Ref. No.			Unit rate pricing	Sheet No.	

> **Typical bill description**
>
> **R12 Drainage below ground**
>
> Excavate trench for pipe not exceeding 200mm nominal size; average depth of trench 750mm; backfilling with excavated material; disposal of surplus off site 174m

ref:	description	quant	unit	rate	lab	plt	mat	s/c	nett unit rate
	item details				*analysis*				
	First calculate the rate for excavating and filling trenches								
	Backacter to excavate at 10m³/hr	0.10	hr	17.00		1.70			
	Banksman	0.10	hr	6.50	0.65				
	Backacter to backfill at 8m³/hr	0.12	hr	17.00		2.04			
	Two labourers to backfill	0.24	hr	6.50	1.56				
	Plate compactor	0.12	hr	1.00		0.12			
	EXTRA for taking 25% to tip								
	7.00-(1.56+0.12)	0.25	m³	5.32		1.33			
	Total rate	1	m³		2.21	5.19			7.40
	For a trench 600 x 750mm deep, assume NO earthwork support								
	but a small over-dig of say 300mm x 750mm deep								
	so volume of excavation = 0.90 x 0.75m = 0.67m³								
	Excavate trench	0.67	m³	7.40	1.48	3.47			
	Trim and compact bottom								
	Labourer 10m²/hr x 0.60m wide	0.06	hr	6.50	0.39				
	Plate compactor	0.06	hr	1.00		0.06			
	Rate for drainage excavation	1	m		1.87	3.53			5.40

CB CONSTRUCTION LIMITED   PRICING NOTES

Project		Trade	DRAINAGE PIPEWORK	Date	
Ref. No.			Unit rate pricing	Sheet No.	

R12 Drainage below ground

A. Granular material type A; bed and surround; to 150mm pipe; 450 wide x 400 deep 174m

B. Clay pipework with flexible joints; in trenches; 150mm nominal size 174m

Typical bill description

ref:	description	quant	unit	rate	lab	plt	mat	s/c	unit rate
A.	Bed and surround								
	First calculate the rate per m³								
Mat	Type A aggregate as quote	1.65	t	6.00			9.90		
	ADD 20% for consolidation and								
	penetration	0.33	t	6.00			1.98		
Lab	Labour previously priced in backfilling tenches								
Plt	Plant previously priced in backfilling tenches								
	EXTRA for disposal of volume occupied by								
	imported filling:	1	m³	7.00		7.00			
	Total	1	m³			7.00	11.88		
	For a machine-excavated trench the minimum width is usually 600mm								
	So the gross volume per metre is 600x400 = 0.24m³/m								
	[larger pipe diameters merit a reduction for the volume occupied by the pipe itslef]								
	Rate for bed and surround	0.24	m³			1.68	2.85		4.53
B.	150mm pipe in trench								
Mat	Pipe from price list with 15% discount	1	m	5.80			5.80		
	Waste 5%	0.05	m	5.80			0.29		
Lab	Drainlayer	0.3	hr	6.50	1.95				
Plt	This method assumes that the excavator								
	can be employed elsewhere when the								
	pipes are laid								
	Rate for 150mm pipe	1	m		1.95		6.09		8.04

| PRICING INFORMATION |  | DRAINAGE TRENCHES |

Width of trench (m)	Pipe sizes (mm)								
	100	150	225	300	375	450	525	600	750
depth ne 1.50m	0.60	0.60	0.70	0.70	0.90	1.10	1.20	1.30	1.45
1.75-2.75m	0.70	0.70	0.80	0.80	1.00	1.20	1.30	1.40	1.60
3.00-4.00m	0.80	0.80	0.90	0.90	1.10	1.30	1.40	1.50	1.70

Note : These are average widths, the actual widths of drainage trenches will depend on :-

    1. The nature of the ground

    2. The method of support to the sides of trenches

    3. The width of bucket if dug by machine

Outputs for drainage gang excavating and laying drains incl backfill	Pipe sizes (mm)								
	100	150	225	300	375	450	525	600	750
Outputs for a drainage gang of one machine, two drainlayers & one labourer (m/day)									
depth (m) 0.50	35	34	33	30	26	25	23		
0.75	34	33	32	29	25	23	21	18	
1.00	32	31	30	28	23	21	19	16	13
1.25	30	29	27	24	20	18	17	14	12
1.50	26	25	24	20	16	15	14	12	11
1.75	21	20	20	17	14	13	12	11	10
2.00	18	17	16	15	12	11	10	9	8
2.25	16	15	14	13	11	10	9	9	7
2.50	15	14	13	12	10	9	8	8	7
2.75	13	12	11	11	9	8	7	7	6
3.00	12	11	10	10	8	7	6	6	6
3.25	10	10	8	8	6	5	5	5	5
3.50	9	9	8	8	6	5	5	5	5
3.75	8	8	7	7	5	5	5	5	4
4.00	7	7	6	6	5	4	4	4	4

Note : These are average production rates, the actual outputs will depend on :-

    1. The nature of the ground

    2. The method of support to the sides of trenches

    3. The length of drainage runs and location

If trenches need to be supported, the cost of hiring trench sheets can be added to the drainage gang rate.

For 20m of trench with both sides supported there would be (20x2)/.33 = 122 sheets (330mm wide) required

With a typical hire rate of £0.60 per week for a 2400mm long sheet, the daily rate would be : 122x0.60/5 = £14.65 per day

The use of trench supports would lead to a reduced daily output by the drainage gang.

CB CONSTRUCTION LIMITED   PRICING NOTES

Project		Trade	DRAINAGE EXCAVATION	Date	
Ref. No.			Operational pricing	Sheet No.	

Typical bill description	CLASS I PIPEWORK - PIPES I112 Clay pipes 150mm dia in trenches, across farmland runs S2-s12; depth not exceeding 1.5m .... 302m L331 150mm bed and surround with 14mm single sized granular material to 150mm dia pipes 302m

| item details | | | | | analysis | | | | nett |
ref:	description	quant	unit	rate	lab	plt	mat	s/c	unit rate
	For excavating a drain run including pipe, bed and surround								
	assume 25m can be completed in one day by the following gang :								
Plt	Backacter	8.50	hr	17.00		144.50			
	Road tipper	2.00	hr	18.00		36.00			
	Plate compacter	8.50	hr	1.00		8.50			
	Trench sheets (see below)	1.00	day	18.24		18.24			
Lab	Labourers (3nr)	25.50	hr	6.50	165.75				
Mat	Pipe from price list with 15% discount	25	m	5.80			145.00		
	Waste 5%	1.25	m	5.80			7.25		
	14mm stone								
	25m x 0.60m x 0.45m x 2.10t/m³	14.175	t	6.10			86.47		
	Rate for one day's work	25	m		165.75	207.24	238.72		611.71
	Rate for drainage (+25)	1	m		6.63	8.29	9.55		24.47
	Trench sheets for 25m = (25x2)/0.33 = 152								
	At £0.60 per week to hire								
	Daily hire rate would be : 152x0.60+5 = £18.24								

PRICING INFORMATION		DRAINAGE MANHOLES

SMM7 NOTES	CESMM3 NOTES
Work section R12 **Drainage below ground**	**CLASS K** **PIPEWORK - MANHOLES AND PIPEWORK** **ANCILLARIES**
Excavation, concrete, formwork, reinforcement, brickwork and rendered coatings measured in accordance with the rules of the relevant work section of SMM7	Items for manholes shall be deemed to include: excavation, disposal, backfilling, upholding sides, concrete work, reinforcement, formwork, brickwork, metalwork, pipework inc backdrops
	Manholes enumerated depending on form of construction
Covers, step irons, channels, benching and building in pipes are enumerated	.... and depths which are given in 1.5m stages
	Depths measured from tops of covers to tops of base slabs
	Types and loading duties of covers are given
	Hand dig is identified in items
	Hard dig given in CLASS L
	Pipe valves measured in CLASS J.

Typical spreadsheet approach to pricing manholes for a specific (civils) project

Description	unit	mat rate	lab/plt rate	1200 ne 1.50m A Mat	Lab/Plt	1350 ne 1.50m A Mat	Lab/Plt	1500 1.50-2.00m B Mat	Lab/Plt
Excavation	m3		5.00		24.20		27.61		31.25
Disposal	m3		7.00		10.08		12.76		15.75
Backfilling	m3		3.50		11.90		12.95		14.00
Side support	m2	1.00	1.00	19.36	19.36	22.09	22.09	25.00	25.00
Surface prep	m2		0.60		0.86		1.09		1.35
Blinding	m3	48.00	19.00	4.84	1.92	6.12	2.42	7.56	2.99
Base slab	m3	42.00	14.00	12.10	4.03	15.31	5.10	18.90	6.30
Surround	m3	44.00	9.00	42.24	8.64	47.52	9.72	52.80	10.80
Formwork	m2		6.00		28.80		32.40		36.00
Chamber rings	nr			75.00	36.00	105.00	48.00	120.00	75.00
Shaft rings	nr								
Reducer slabs	nr								
Cover slabs	nr			60.00	14.00	80.00	17.00	110.00	22.00
Cover	nr			85.00	12.00	85.00	12.00	45.00	10.00
Benching	m3	55.00	35.00	15.84	10.08	20.05	12.76	24.75	15.75
Channels	item			24.00	12.00	24.00	12.00	33.00	15.00
Brickwork	m2	23.00	32.00	10.58	14.72	10.58	14.72	10.58	14.72
Sundries	item								
			totals	348.95	208.59	415.67	242.63	447.59	295.91

Example of preliminaries build up for a three weeks project with an anticipated value of £30,000

CB CONSTRUCTION LIMITED   SMALL WORKS PRELIMINARIES

Project: Plant foundations at Stansford Quarry			Number: T377		Date: July 1994	

General Expenses					Breakdown of totals						
Description	Quant	Unit	Rate	Total	lab	plt	mat	s/c		staff	o/head
1 Supervision											
a Agent	3	wks	600	1800						1800	
b Setting out Engineer	1	wks	450	450						450	
c QS/Estimator	1	wks	550	550						550	
d General Foreman		wks									
			total	2800						2800	
2 Site Establishment											
a Notice board	1	nr	25	25		25					
b Hardstanding		m2									
c Office	3	wks	25	75		75					
d Messroom		wks									
e Stores	3	wks	25	75		75					
f Toilets	3	wks	15	45		45					
g Foundations	1	item	25	25			25				
h Transport/crane		item									
j Install temp elec		item									
k Install temp water	1	item	75	75							75
l Install temp drains		item									
m Office furniture/equipment	1	item	35	35							35
n Fence around compound		m									
	0		total	355		220	25				110
3 Site Running Expenses	0										
a Elec charges		qu									
b Water charges		%									
c Telephone charges	3	wks	35	105							105
e Postage & stationery	3	wks	35	105							105
f Photocopying & printing	3	wks	30	90							90
g Setting out consumables	1	sum	40	40							40
h Protective clothing	1	sum	35	35							35
j Photographs	1	sum	30	30							30
k Materials testing	6	tst	5	30							30
l Consumables	3	wks	20	60							60
			total	495							495
4 General Labour											
a Set-up/clear site	2	wks	200	400	400						
b Distribution/plant driver		wks									
			total	400	400						
5 Plant											
a Forklift		days									
b Dumper		wks									
c Crane	2	visits	180	360		360					
d Compressor & tools		wks									
e Concrete pump		days									
f Mixer		wks									
g Water pump	3	wks	100	300		300					
h Level	3	wks	20	60		60					
j Theodolite	1	wks	30	30		30					
k Skips	2	nr	40	80		80					
l On/off site transport	2	nr	40	80		80					
m Minor plant (% of labour)	2	%	3700	74		74					
			total	984		984					
6 Temporary works											
a Structures		sum									
b Scaffolding		sum									
c Access road/hardstanding	100	m2	3.00	300			300				
			total	300			300				
8 Insurances											
a Insurances	0.4	%	30000	120							120
			total	120							120
Preliminaries Total			£	5454	400	1204	325			2800	725

Sub-contractors and nominated suppliers

Introduction

Sub-contractors can be classified into two main categories: nominated and domestic. The way in which a sub-contractor's quotation is incorporated into the tender will depend on the contractual relationship of the specialist with the main contractor, and the definitions given in the standard method of measurement used. During the 1980s and early 1990s the use of the formal nomination procedure diminished. It has been replaced by lists of approved sub-contractors given in the tender documents, named sub-contractors where the Intermediate Form is used, and the novation of specialists. (*The word 'novation' means the substitution of a new obligation for an old obligation by the mutual consent of the parties, and is used where the client has already completed negotiations with a sub-contractor or consultant and invites the contractor to enter the agreement.*)

Domestic sub-contractors

The procedures for despatching enquiries to obtain quotations from sub-contractors were given earlier. Great care must be taken when estimating on the basis of sub-contract quotations because the contractor takes responsibility for all the work. It is therefore for all the parties to ensure that quotations are based on accurate and complete information.

All sub-contract quotations should be checked for arithmetical errors and totalled. To compare them on a like-for-like basis the following checks are carried out by the estimator:

1. All the items for that trade should be priced. If there is enough time, the sub-contractor should be asked to provide missing rates, otherwise the estimator needs to insert his own estimate of their value.
2. The rates should be realistic. If a sub-contractor finds that he has made a

163

patent error then he could withdraw his tender or change his offer before it is accepted. It is most important that the sub-contractor should be advised to amend his quotation and tell all the main contractors who have received it.

3. It is sometimes argued (mainly by quantity surveyors) that rates should be consistent throughout the bill of quantities – like items should be priced at similar rates to avoid possible difficulties when valuing variations. Anyone vetting a tender must realize that the cost of similar items may vary depending on quantities, location, timing and so on.

4. The sub-contractor should accept the contract conditions without amend-ment. This will enable the estimator to make fair comparisons between quotations, and avoid any misunderstandings brought about by qualified bids. In practice, quotations are sent with many printed and specific conditions which may conflict with the enquiry documents. These details are often resolved at the negotiation stage.

5. The quotation should be based on the documents which form the main contract. The estimator should accept neither a lump-sum quotation for work which will be valued on the basis of an approximate bill of quantities or a schedule of rates for a plan and specification project. If a sub-contractor has altered the tender documents (in the bill of quantities, for example) there may be a mistake which should be reported to the client so that all the contractors will correct the bill before the tenders are submitted.

Quotations from specialists often need careful comparison using a standard form. The example sheet shown in Fig. 12.1 can be used to compare 'supply and fix' sub-contractors with labour-only contractors; the difference between the two is usually the cost of materials.

A computer can be a great help in comparing sub-contractors' quotations. Spreadsheet software is particularly useful for listing, and comparing rates, and provides a mathematical check. The spreadsheet method also allows rates to be adjusted before they are put in the estimate. Computer-aided estimating packages offer more powerful facilities, in particular:

1. The software will prompt the estimator by showing items which the sub-contractor should have priced.
2. Average rates can be inserted automatically when one of the sub-contractors fails to price an item.
3. The chosen list of rates can be incorporated into the estimate at the touch of a button.

This assumes that the estimator wants to insert the rates as they stand. In some cases a lump sum may be added to certain rates or a percentage may be applied to others. The main reasons for changing the sub-contractor's rates are:

CB CONSTRUCTION LIMITED — Sub-contract comparison sheet

Project: LIFE BOAT STATION Ref.No. T384 Date 7.94

Page	Item	Description	Quant	Unit	S/C PRELUDE STONE rate	£	S/C rate	£	S/C DEANSTONE rate	£	S/C rate	£	LOSC CASTLE rate	£	MATS RENARD QUARREY Basic	Sund	Waste 2½%	Rate	£
4|13	C	COPING 550 x 175	23	m	233.30	5366			265.00	6095			36.04	829	183.30	1.20	4.61	189.11	4350
	D	MITRED ANGLE	8	Nr	28.00	214			inc	–			inc	–	18.00	–	0.45	18.45	148
4|10	E	JAMB 120 x 160 x 1114	186	Nr	107.00	19902			125.50	23343			72.10	13411	56.50	1.00	1.44	58.94	10963
	F	JAMB 135 x 160 x 1114	38	Nr	110.00	4180			129.40	4917			73.20	2782	57.30	1.00	1.46	59.76	2271
	G	CILL 215 x 175 x 839	2	Nr	165.10	330			173.60	347			90.10	180	102.21	1.80	2.60	106.61	203
	H	CILL 215 x 175 x 1014	94	Nr	186.58	17727			195.30	18358			90.10	8469	125.70	1.80	3.19	130.69	12285
	J	HEAD 230 x 225 x 1014	15	Nr	177.96	2669			235.00	3525			144.16	2162	115.08	1.20	2.91	119.19	1788
	K	MULLION	3	Nr	186.25	559			195.30	586			144.16	432	110.00	1.50	2.79	114.29	343
4|11	A	STAINLESS STEEL DOWELS	333	Nr	5.40	1798			4.85	1615			3.20	1066	1.80	0.80	0.07	2.67	889
						£ 62755				£ 58786				29331					£ 33240
					TRANSFER									33240					
					TO ESTIMATE								£ 62571						

Fig 12.1 *Example of a sub-contract comparison sheet*

1. The sub-contractor might need specific builder's work not measured elsewhere, drilling holes through the building fabric being the most common example.
2. An estimator might decide to add certain attendances to the measured rate, such as scaffolding for an industrial door installer.
3. There may be specific trade requirements which are customarily provided by the contractor. For example, a piling firm may ask for surplus soil to be removed by the main contractor, and a plasterer often expects the free use of a mechanical mixer.
4. A margin for overheads and profit could be added to all or some of the rates; either because the contractor wishes to spread his overheads through the bill or for tactical reasons, such as work which appears to be undermeasured.

There is a slight danger that adding attendances and margin to rates may confuse site surveyors or buyers. This should not be a problem if staff understand the distinction between net allowances and the rates given in the client's (gross) bill. The former are target rates for buying materials and services, the latter being the value the contractor will be paid for his services. To keep matters simple, it is customary to deal with the attendances and overheads in the assessment of site overheads.

Figure 12.2 shows the attendances to be provided by the main contractor, without charge. These attendances are defined in the form of sub-contract, and are normally priced in the project overheads. There will be some sub-contractors, of course, who will need more than others. A cladding contractor, for example, will need a considerable amount of safety and access equipment whereas a plasterer may only need a small mixer and a supply of clean water.

Whenever a particular sub-contractor is used in the tender, an entry should be made on the summary of domestic sub-contracts (see Fig. 12.3). If a lower quotation is received later in the tender period, an adjustment can be made on this form and carried to the tender analysis reports presented to management at the adjudication meeting.

Nominated sub-contractors

The nomination procedure suffers from an elaborate set of conditions in the JCT80 contracts which has had the effect of turning people away from the practice of nominating specialist sub-contractors and suppliers. The general conditions for government contracts, GC/Works/1(Edition 3), is much simpler; clause 63 (Nominations) starts with the following declaration:

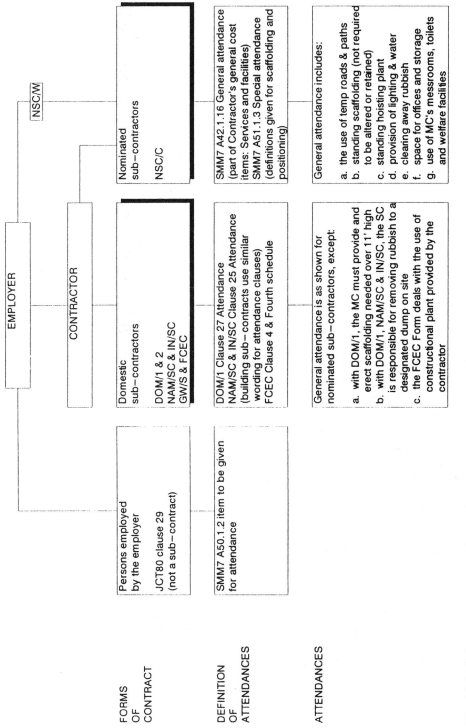

Fig 12.2 Sub-contract types and attendances

CB CONSTRUCTION LIMITED	DOMESTIC SUB-CONTRACTORS SUMMARY					Project	Fast Transport	
						Ref.No.	T354	Date	11.7.94

Ref.	Trade	Company	Quotation	Discount offered (%)	Net amount	Firm price Allowance	Alternative quotations		
							Company	Net amount	Saving
S1	Roof covering	Beaufort Roofing	17672	2.5	17230	nil			
S2	Windows	Valley Fabrications	30641	2.5	29875	1225	Archiglass	27550	2325
S3	Plumbing	Consort	4550	nil	4550	nil			
S4	Plastering & partitions	Swift Services	57990	nil	57990	nil	Oscar Finishes	46760	11230
S5	Joinery	Projoin Site Services	41900	nil	41900	1935	L.P.Monk	38450	3450
S6	Suspended ceilings	Hill Systems	19882	2.5	19882	nil			
S7	Painting	Tudor Decorations	12659	2.5	12343	nil			
S8	Floor coverings	Freedom Finishes	12615	2.5	12300	nil			
S9	Electrical installation	Comech Engineering	35887	2.5	34990	nil	Beta Technologies	22860	12130
S10	Mechanical installation	Comech Engineering	25667	2.5	25025	nil			
S11	Surfacing	W. Smith Contracting	11800	nil	12450	nil			
S12	Scaffolding	CCG Scaffolding	see prelims						
	Totals			£	268535	3160		£	29135

Fig **12.3** *Typical summary form for domestic sub-contractors in a tender*

A nominated subcontractor or supplier means a person with whom the Contractor is required to enter into a contract for the execution of work or the supply of Things designated as 'Prime Cost' or 'PC' items. This requirement may be specified in the contract documents or in any direction or Instruction given under the Contract.

The standard method of measurement gives the rules for items to be included for nominated sub-contractors (SMM7 rule A51) and nominated suppliers (SMM7 rule A52). The following information must be given for each nominated sub-contractor:

1. The nature and construction of the work.
2. A statement of how and where the work is to be fixed.
3. Quantities which indicate the scope of the work.
4. Any employer's limitations affecting the method or timing of the works.
5. A prime cost sum.
6. General attendance item in accordance with Section A42.
7. An item for main contractor's profit, to be shown as a percentage.
8. Details of special attendance required by the sub-contractor.

SMM7 lists some of the special attendances which might be required and makes it clear that special scaffolding is that which is needed as well as the contractor's standing scaffolding provided for other trades. The companion document 'SMM7 Measurement Code' goes further, suggesting that the bill item for special scaffolding should be accompanied by dimensions. One of the biggest problems for estimators is the common practice of quantity surveyors who merely list all the general items given in SMM7 against every nomination. Again the Measurement Code warns against this practice and states that where adequate information cannot be provided a provisional sum should be used. Furthermore, items for positioning should state the expected weight, location and size of the components to be positioned.

The estimator's task is to make sure that the PC sum gets included in the bill calculations, adding a percentage for profit to the PC sum if it is wanted, considering the effect of the work on the programme, and assessing the costs of general and special attendances. Many contractors are reluctant to insert a figure for profit against the bill item because it might be interpreted as representing the contractor's profit on all the work.

SMM7 defines general attendance in Section A42 coverage rule C3 (see Fig. 12.2). The estimator usually provides for the cost of these attendances in his evaluation of the project overheads since most of the facilities are common to other trades. Special attendances, on the other hand, need to be priced separately either in the measured bill or as items in the project overhead calculations. If the special attendances are properly described the estimator can price the work; but where general descriptions are used he has great difficulty assessing his obligations. A typical example is a prime cost sum for piling where the type of pile is not given and the special attendances state the need for 'positioning'. SMM7 tells us that positioning means unloading, distributing, hoisting and placing in position. Does the piling contractor expect the contractor to do all this for him?

Nominated suppliers

SMM7 and the standard forms of contract state that a nominated supplier is identified in the tender documents as a prime cost sum. A separate item is given for the contractor to add his profit. JCT80 says that the nominated supplier shall allow the contractor a discount for cash of 5%. Many contractors take out this discount before adjudication so the meeting can consider the net costs before considering the mark-up required to convert the estimate to a tender. A building estimator must be careful in dealing with GC/Works 1/Edition 3 because clause 63(3) states that the sum paid to the contractor by the authority, for all nominations, is the prime cost after the deduction of 'all discounts rebates or allowances'. No such problem arises with ICE 6th Edition where the contractor can keep the discount obtainable for prompt payment to firms nominated for services or the supply of goods.

Fixing materials provided by nominated suppliers is measured in the appropriate part of the bill of quantities. GC/Works 1/Edition 3 points out that rates for fixing must include unloading, getting-in, unpacking, return of empties and other incidental expenses. A similar definition was given in SMM6 and was added to SMM7 with amendment No. 2 (New Coverage Rule C1 for section A52.1).

Fluctuations

The fluctuations calculation

Introduction

In the majority of cases – all but the largest of projects – contractors must allow for changes in costs which occur during the construction phase. The amount which the estimator adds to the estimate for inflation is a guess, calculated after an examination of price trends over the previous few years, and discussions with

171

suppliers and sub-contractors, in an attempt to predict future trends. If a contract is likely to last for several years then the employer will request a tender based on current prices and any changes will be reimbursed using the methods defined in the contract.

The standard forms of contract have terms for either a *firm* or *fluctuating* price. A firm price is one which will not be varied for changes in the cost of resources, although labour tax fluctuations are usually reimbursed by the employer. In a fluctuating-price tender the price is agreed before the job starts but the contract sum can be adjusted for changes in the costs of resources. An estimator needs to understand the clauses dealing with fluctuations so that he can tell the sub-contractors of the risks and calculate his own forecast of increased costs.

Standard fluctuations clauses

The JCT Standard Form of Building Contract lists three options under clause 37 for dealing with fluctuations. The choice is noted in the Appendix as one of three clauses 38, 39 or 40. The estimator should find this information in the preliminaries section of the bill of quantities. If a method has not been chosen then clause 38 shall apply (firm price). Since only one option is required, fluctuation clauses are published separately from the rest of the contract conditions.

There are, in fact, four possibilities for dealing with fluctuations:

1. No clause – not recognized by the JCT contracts but would produce a firm price regardless of statutory or other changes.
2. Clause 38 – is the firm price alternative which allows some statutory changes such as levies, taxes and contributions.
3. Clause 39 – is the full fluctuation option allowing variations in prices of labour, and materials using a basic list.
4. Clause 40 – is the 'Formula Method' where changes in costs are calculated by applying average indices prepared by national bodies published in monthly bulletins.

Clause 39 is the traditional method for the recovery of increases in the costs of employing labour and buying materials, but is seldom used today. It is an attempt to calculate the actual increases or decreases in costs incurred by the contractor and his sub-contractors. This method leads to a great deal of work for the quantity surveyors because the actual costs of construction must be compared with those at the date of tender. No increases are allowed for:

1. Overheads and profit.
2. Site supervision.
3. Site establishment costs.
4. Plant and temporary works.

The contract allows the contractor to enter a percentage addition in the Appendix for some of the costs which are not recoverable.

Clause 40 (the price adjustment formulae option) is not based on the actual cost changes as in Clause 39; instead it uses the changes in indices published by the DOE which are published monthly. There is therefore no need for a basic list of material prices and the administrative work is reduced. There are 49 work categories covering general building work and many common specialist activities. The PQS should assign the bill items to work categories so that the contractor can assess the way in which increases will be dealt with when the work is valued. Fluctuations are not calculated for the following:

1. Credit for demolition materials.
2. Unfixed materials on site.
3. Plant paid for on a daywork basis (labour is reimbursed at the rate current when the work is carried out).
4. Claims (normally calculated at full value).
5. A non-adjustable element which is deducted from the increased costs payable under the Local Authorities editions of JCT 80. The deduction, which is normally 10%, is made because it could be argued that contractors should not receive an addition on overheads and profit.

The Joint Contracts Tribunal has produced similar conditions for dealing with fluctuations under the Intermediate Form of Building Contract. Clause 4.9 states that the contract sum will be adjusted for contribution, levy or tax matters unless the price adjustment formulae method is given in the Appendix to the contract. There are no fluctuations rules in the Agreement for Minor Building Works because the contract is for work of short duration. The usual clause for contribution, levy and tax changes is written into the contract (clause 4.5) but can be deleted if the contract period is short.

Works contractors engaged by management contractors under the terms of the JCT Works Contract conditions are reimbursed using one of three methods which are similar to the JCT 80 clauses 38, 39 and 40. The JCT Management Contract itself does not have provisions for fluctuations, presumably because the management contractor is paid the prime cost of the work.

The ICE Conditions of Contract include a supplementary clause for Contract

Price Fluctuations, attached in looseleaf form. Again the most common arrangement is for a firm-price contract without fluctuations but including (under clause 69) provision for statutory labour taxes, levies or contributions. The Central Government form GC/Works/1 uses a similar arrangement for labour tax matters and supplementary conditions for fluctuations using the NEDO formulae method for recovery.

Calculation of non-recoverable increases

In broad terms the calculation shown below is necessary on every contract, including those with fluctuations clauses, because there is usually a shortfall in recovery of increased costs:

A. Forecast increases in costs of resources.
B. Forecast amount recoverable.
C. *Add* non-recoverable element (A − B) to tender.

Now that analytical estimating is widely used by estimators, the fluctuations calculations can be dealt with after the bill of quantities has been rated – all the resources can be examined separately and a forecast of changes can be made. On the other hand, if the labour element is not known, a wage increase adjustment can be made to the all-in rate before pricing begins.

The tender programme is an invaluable aid in forecasting cost increases, not only for the construction phase but also to calculate the effect of the time between date of tender and start on site. The period for which the tender is to remain open for acceptance should be as short as possible if the employer wants to receive an economical price for the work. Figure 13.1 shows the use of a tender programme in assessing fluctuations. No labour increase is expected in this example until the following June which is after completing the project. If the project started in March (Fig. 13.2) there would be a June increase in labour rates but no more costs for staff who have their salaries increased in January.

It can be difficult to forecast changes in costs accurately. A reasonable estimate can be made, however, if individual resources are treated separately, as follows:

Labour The forecast of labour cost increases is the most predictable part of an estimate because wages change on the same date each year and the increase follows political and economic trends. Historical data can be plotted on a graph if a longer-term view is needed. For labour-only sub-contractors changes are difficult to anticipate and at times of recession can decrease more quickly.

ESTIMATE OF FLUCTUATIONS

	LAB	PLT	MAT	DOM S/C	STAFF		
1 Pre-construction						< TENDER	
2 Excavation & filling						< AWARD	
3 Foundations fmwk / conc							
4 Underslab drainage							
5 Concrete grd floors			429			mat	£10730 @ 4%
6 Columns fmwk / conc			62			mat	£2055 @ 3%
7 Floors & beams fmwk / conc			333			mat	£9670 @ 5%
8 External walls			2112			mat	£42250 @ 5%
9 Roof timbers			275			mat	£5500 @ 5%
10 Roof covering			58	856		mat / s/c	£1170 @ 5% / £17125 @ 5%
11 Windows				975		s/c	£19500 @ 5%
12 Services 1st fix				FIRM PRICE			
13 Plasterwork & partitions				FIRM PRICE			
14 Joinery			342	930		mat / s/c	£4885 @ 7% / £15500 @ 6%
15 Ceilings			86	850		mat / s/c	£1225 @ 7% / £14175 @ 6%
16 Services 2nd fix				FIRM PRICE			
17 Painting				FIRM PRICE			
18 Floor coverings				711		s/c	£9480 @ 7.5%
19 External works & drainage			800			mat	£16000 @ 5%
20 Prelims							
– Agent					}	SALARIES AFTER ANNUAL REVIEW	
– Engineer					630		
– Foreman					}	£9000 @ 7%	
– General labour					}	£1000 @ 6%	
– Forklift		60					
Fluctuations TOTALS	60	60	4497	4322	630		

<STAFF PLUS 7%

Fig 13.1 Calculation of fluctuations for a project starting in September

ESTIMATE OF FLUCTUATIONS

#	Item	LAB	PLT	MAT	DOM S/C	STAFF	
1	Pre-construction						
2	Excavation & filling						< TENDER ... < AWARD
3	Foundations fmwk / conc						
4	Underslab drainage						
5	Concrete grd floors			429			
6	Columns fmwk / conc			62			
7	Floors & beams fmwk / conc			333			
8	External walls			2112			
9	Roof timbers	93		275			lab £1550 @ 6%
10	Roof covering	75		58	856		lab £1245 @ 6%
11	Windows				975		
12	Services 1st fix				FIRM PRICE		
13	Plasterwork & partitions				FIRM PRICE		
14	Joinery	166		342	930		lab £2770 @ 6%
15	Ceilings	63		86	850		lab £1055 @ 6%
16	Services 2nd fix				FIRM PRICE		
17	Painting				FIRM PRICE		
18	Floor coverings				711		
19	External works & drainage	886		800			lab £14770 @ 6%
20	Prelims						
	– Agent					}	
	– Engineer					} nil	
	– Foreman					}	
	– General labour	286					lab £4770 @ 6%
	– Forklift		60				
	Fluctuations TOTALS	1569	60	4497	4322		

estimate of labour increase = 6%

Timeline months: January, February, March, April, May, June, July, August, September

Fig 13.2 Calculation of fluctuations for a project starting in March (figures for materials, plant and sub-contractors brought forward from Fig. 13.1)

There are three ways to calculate expected labour costs throughout the currency of the contract:

1. Adjust the all-in rate before pricing the bill so all labour is priced on an average rate. This can be done as follows:

May–June	8 weeks at £5.60	£44.80
July–Sept.	12 weeks at £6.10	£73.20
Total		£118.00
Average rate	118.00/20=	£5.90

2. Increase the total labour costs by a percentage based on the approximate amount of work to be carried out before and after a wage increase.
3. Increase only the trades which are working after the wage increase by examining the tender programme.

Staff   The cost of staff can be split between the amounts before and after the annual review. A simple percentage can then be added to the total salaries after the review.

Materials and plant   There are two methods commonly used to assess the increased costs of materials and plant:

1. The estimator could assume a constant increase over the time that resources are purchased or hired and determine the increases up to the mean purchase date. As a simple example, if the estimator uses a constant increase of 1% a month for a 12-month contract he might add 6% to the total value of materials.
2. A more accurate assessment can be made by looking at each material or item of plant abstracted in the estimator's summary. A separate decision can then be made for each resource for likely increases and their timing with respect to the programme. If concrete prices are due to increase, for example, during the last third of a building project programme there may be no need to add for inflation.

Sub-contractors   All enquiries to sub-contractors must clearly state the rules for fluctuations. The estimator must check their quotations for compliance with the conditions. For some trades, such as surfacing, the estimator may have difficulty obtaining offers which fully comply with a request for a firm-price tender. He will identify the problem and adjust the sub-contract value when he completes the domestic sub-contract register; an example is given on page 54 of the COEP. On the other hand, the firm price adjustments could be made on a sub-contract summary sheet (see Chapter 12) which lists all the sub-contract values used in the tender; there are no examples of this summary in the COEP.

For a *firm-price* tender the estimator will add all the increases to his estimate; although in a strongly competitive market he may assume that certain price increases can be avoided or negotiated away. For a contract which is based on a *formula method* the estimator must add the non-adjustable element to his price. This is an amount taken from increased costs where the Local Authorities edition of JCT 80 is used. The deduction, which is normally 10% of the increases, is made because it could be argued that contractors should not receive an addition on overheads and profit.

Provisional sums and dayworks

Introduction

Contractors have traditionally added the amounts for provisional sums and dayworks after the profit margin has been calculated. This is because, when provisional sums and dayworks are valued during the contract, the contractor receives adjustments for overheads and profit. This changed in 1988 when SMM7 introduced the use of provisional sums for *defined work* and the JCT standard contract forms were amended accordingly.

Provisional sums for undefined work

Where the employer identifies that there is likely to be extra work for which there is no information at tender stage or it cannot be measured using the standard method of measurement, a provisional sum can be provided in the bill of quantities. The sum is spent at the direction of the architect (or engineer, ICE 6th Edition) and the work is valued in accordance with the valuation rules. There are two kinds of undefined provisional sum; a contingency sum which is for work which cannot be identified at tender stage, usually for unforeseen circumstances, and sums for specific items the extent of which is not known, such as more landscaping to a courtyard which has not been agreed with the client. The contractor adds these provisional sums to his tender after he has calculated his preliminaries and profit margin. SMM7 makes it clear that the contractor is entitled to any reasonable allowance for programming, planning and preliminaries. This is not just a financial compensation. JCT80 (clause 25.4.5) gives the expenditure of a provisional sum as a relevant event which may lead to a claim for an extension of time. Taken literally, this means that the estimator does not include provisional sums when planning the work.

Provisional sums for defined work

SMM7 recognizes that there are certain items of work which cannot be measured using the standard method but could be taken into account by the contractor when he draws up his programme and calculates his overheads and profit. Simple examples would be providing a concrete access ramp for wheelchairs, or intumescent paint to roof trusses.

Contractors must be given more information about work in this category so that all the temporary works and overheads can be calculated. The question is, how much information must be given in the bill of quantities for a provisional sum for defined work? SMM7 states that the following must be provided:

1. The nature of the work.
2. How and where it is to be fixed.
3. Quantities showing the scope and extent of the work.
4. Limitations on method, sequence and timing.

Estimators have experienced problems with bills of quantities with provisional sums for defined work where the full extent of the temporary works is not clear. A typical case would be a provisional sum to replace defective windows in a multi-storey building. What assumptions should the contractor make for scaffolding? The defective windows could be found at high level, lower levels or throughout the building.

Dayworks

Construction contracts often involve changes from the original scheme. The term *variation* means alteration of the design, quality or quantity of the works, and can include changes in sequence or timing of the works. Where a variation occurs, the cost of the original work is deducted and new work is measured and priced by the quantity surveyor or engineer. The value of variations is determined according to the rules set out in the conditions of contract. The first method is to be by measurement using bill rates or a fair allowance added to the bill rates or by fair rates and prices. Where the work cannot be valued by measurement, it may be valued on a daywork basis, provided it is incidental to contract work. In practice, contractors are often asked to attach a value to a variation before the varied work is started and in some cases a term is incorporated into the contract so that agreement is reached before the work is carried out.

The daywork charges are calculated using the definitions prepared for building works by the RISC/BEC and civil engineering work by the FCEC. The JCT Standard Form of Building Contract (1980 edition) states that where the valuation relates to additional or substituted work which cannot be properly valued by measurement the valuation shall comprise:

the prime cost of such work (calculated in accordance with the Definition of Prime Cost of Daywork carried out under a Building Contract issued by the Royal Institution of Chartered Surveyors and the Building Employers Confederation which was current at Base Date) together with percentage additions to each section of the prime cost at the rates set out by the Contractor in the Contract Bills; or

where the work is within the province of any specialist trade and the said Institution and the appropriate body representing the employers in that trade have agreed and issued a definition of prime cost of daywork, the prime cost of such work calculated in accordance with that definition which was current at the Base Date together with percentage additions on the prime cost at the rates set out by the Contractor in the Contract Bills.

A footnote states that the RICS has agreement about the definition of prime cost with two specialist trades associations – electrical and heating and ventilating associations. The contractor's daywork rates (or percentages) must take into account the rates required by the sub-contractors used in the tender.

The two industry definitions of prime cost of daywork state that the component parts which make up a prime cost are labour, materials, plant, and supplementary charges in the case of the civil engineering definition. The contractor adds for incidental costs, overheads and profit at tender stage, thus introducing competition into the daywork part of the tender:

1. *Labour* For building works, the hourly base rates for labour are calculated by dividing the annual prime cost of labour by the number of working hours per annum (see Fig. 14.1). The annual prime cost of labour comprises:
   (a) Guaranteed minimum weekly earnings
   (b) Extra payments for skill
   (c) Payments for public holidays
   (d) Employer's National Insurance contributions
   (e) Annual holiday credits
   (f) Contributions to death benefit scheme
   (g) Contribution, levy or tax payable by employer
2. *Materials* The prime cost of materials is the invoice cost after deducting trade discounts, but include cash discounts up to 5%. For civil engineering and government contracts the cash discount kept by the contractor cannot exceed 2.5%.

Calculation of prime cost of labour for daywork		days	hrs		
Working hours per week		39			
Working hours per year			2028		
Annual holidays (enter days)		21	−164		
Public holidays (enter days)		8	−62		
Total working hours per annum			1802		

Calculation for 1991/92		craft	lab	CRAFT	LAB
Guaranteed weekly wage		159.51	135.91		
Annual costs for working hours		46.2 weeks		7369.36	6279.04
Annual costs for public holiday		1.6 weeks		255.22	217.46
Extra for skill/hr	1802	0.10	0.20	180.18	360.36
Employer's Nat Ins (%)		8.60	8.60	671.21	589.69
Training levy (%)		0.25	0.25	19.51	17.14
Annual holidays & death benefits		19.20	19.20	902.40	902.40
Annual prime cost of labour				9397.88	8366.09
Hourly base rate	1802 hours			£5.22	£4.64

Overhead	Rates incl overheads	
%	CRAFT	LAB
20	6.26	5.57
30	6.78	6.04
40	7.30	6.50
50	7.82	6.96
60	8.35	7.43
70	8.87	7.89
80	9.39	8.36
90	9.91	8.82
100	10.43	9.29
110	10.95	9.75
120	11.47	10.22
130	12.00	10.68

Fig 14.1 *Estimator's spreadsheet for calculating the prime cost of building labour for daywork 1992/1993*

3. *Plant* The definitions include schedules for plant charges. They relate to plant already on site and the rates include the cost of fuel, maintenance and all consumables. Drivers and attendants are dealt with under the labour section. In the civil engineering definition the rates provide for head office charges and profit.

Overheads and Profit

The anticipated value of daywork is included in a bill of quantities as provisional sums for labour, materials, plant, and supplementary charges in the case of civil engineering work. The contractor is invited to add a percentage to each section for incidental costs, overheads and profit. The civil engineering definition states the percentage additions; but the contractor still has the opportunity to add or deduct from the percentages given. Overheads and profit are calculated as follows:

1. *Labour* The estimator must calculate the hourly base rate (see Fig. 14.1) and compare it with an 'all-in' rate which includes overheads and profit. The rate or percentage in a tender is an average for all types of labour regardless of trade or degree of supervision because the estimator has no idea what extra work will arise. One solution is to look at a similar job and compare the net cost of measured work with the tender sum. It might be found that the total mark-up including project overheads was 30%. If the all-in rate for the current estimate is £6.44 then the gross all-in rate would be £6.44 × 1.30 = £8.37/hour. Fig. 14.1 shows the comparable hourly base rate to be £5.22. This would suggest a percentage to be added to the hourly base rate of $(8.37 - 5.22)/5.22 \times 100 = 60\%$.

   Fig. 14.2 shows the incidental costs, overheads and profit items listed in section 6 of the RICS definition and highlights the items which need to be added to the all-in rate. The answer could also be found by comparing the all-in rate used in the estimate with the daywork base rate, as follows:

Hourly base rate for labour (Fig. 14.1)	£5.22
All-in hourly rate (Chapter 10, Fig. 10.1)	£6.44
% increase	23
(a) Head office charges	5
(b) Site supervision	8
(j) Insurances	2
(o) Scaffolding	2
(p) Temporary works and protection	2
(q) Other liabilities	10
(s) Profit	5
Total percentage to add to standard rate	57%

Abstract from Section 6 of the RICS/BEC Definition

Item	Included in all-in rate	Not included
(a) Head office charges	—	×
(b) Site staff including site supervision	—	×
(c) The additional cost of overtime (other than authorized)	×	—
(d) Time lost due to inclement weather	×	—
(e) Additional bonuses and incentive schemes	×	—
(f) Apprentices' study time	—	×
(g) Subsistence and periodic allowances	—	×
(h) Fares and travelling allowances	×	—
(i) Sick pay or insurances in respect thereof	×	—
(j) Third party and employer's liability insurance	—	×
(k) Liability for redundancy payments	—	×
(l) Employer's National Insurance contributions	×	—
(m) Tool allowances	×	—
(n) Use, repair and sharpening of non-mechanical tools	—	×
(o) Use of erected scaffolding, staging, trestles and the like	—	×
(p) Use of tarpaulins, protective clothing, artificial lighting, safety and welfare facilities storage and the like that may be available on site	—	×
(q) Any variation to basic rates required by the Contractor in cases where the building contract provides for the use of a specified schedule of basic plant charges (to the extent that no other provision is made for such variation).	Applies to plant section	
(r) All other liabilities and obligations whatsoever not specifically referred to in this section nor chargeable under any other section	—	×
(s) Profit	—	×

Fig 14.2 *Items to be added to the 'all-in' rates for labour*

The first calculation produced an answer of 60% and the second 57%. So why do some contractors want 110% and some specialists ask for over 150%? The four reasons for higher percentages, given below, go some way towards answering this question but would not be valid if dayworks were only *incidental* to and not a significant part of contract works:

(i)   The rates paid by contractors to labour-only sub-contractors are often higher than the all-in rate for direct employees, and, when the work is plentiful, the market rate for labour can be substantially higher.

(ii) Introducing variations into a normal work sequence can have a harmful effect on other work and the attitude of the workforce particularly when changes make it difficult to earn the expected bonuses.

(iii) The rate quoted by the main contractor, for building work, must include the possibility of work being carried out by specialist sub-contractors; their operatives may be earning higher rates of pay which are not recognized by the agreed definition, and specialist fitters often want a subsistence allowance while working away from home.

(iv) Contractors, again in the building industry, do not add the full percentages to materials and plant because they assume that labour is the major element of daywork; this means that the labour percentage must carry the overheads which have not been added to materials and plant.

2. *Materials* Most of the items for incidental costs, overheads and profit, listed in Fig. 14.2, do not apply to materials. Contractors therefore add a small percentage (usually between 10% and 15%) to cover head office overheads (a) and profit (s). The costs incurred in unloading and transporting materials around the site would be fully recoverable.

3. *Plant* The rates for plant are given in the definition as provided by the contract. In building work, using the RICS/BEC definition, it is important to identify when the schedule of plant rates was produced in order to allow an additional percentage if the schedule is out of date. Section 6(q) of the RICS definition provides this opportunity. Schedules of plant charges usually cover a wide range of equipment, and apply to plant already on site. The costs which a contractor can claim are for the use of mechanical plant, transport (if plant is hired specifically for daywork) and non-mechanical equipment (except hand tools) for time employed on daywork. Labour operating plant is dealt with in the labour element.

The estimator must try to assess which are the most likely pieces of equipment to be used and compare the scheduled rates with those quoted by local plant hirers. The allowance for overheads and profit is commonly quoted between 10% and 15%.

Preliminaries

Introduction

The preliminaries bill gives the contractor the opportunity to price project overheads, which are defined in the Code of Estimating Practice (COEP) as 'The cost of administering a project and providing general plant, site staff, facilities and site-based services and other items not included in all in rates'. The standard methods of measurement for civil engineering and building give the general items which should be described in a bill of quantities, in two main parts: the specific requirements of the employer and the facilities which would be provided by the contractor to carry out the work. It could be argued that the latter is not really necessary in a bill of quantities because the contractor must provide general facilities whether they are measured or not. Presumably a simple breakdown of a contractor's general cost items is needed to make a fair valuation of the works during the construction phase. This approach to measurement can lead to duplication of descriptions, because an item may be required by a client and is something which a contractor would normally provide. It is common, in building for example, to find preliminary descriptions for security and protection of the works measured twice.

Pricing preliminaries

For small repetitive works a contractor or sub-contractor may have a scale of overheads which he can apply to a new project. This may be calculated as a percentage of annual costs and adjusted where jobs deviate from the norm. Typically the overheads for small houses and extensions may be 15%, and where sub-contractors have facilities provided by main-contractors the figure would be nearer 10%.

Traditionally, in building, estimators have allowed for attendant labour, non-mechanical plant and certain items of mechanical plant in the rates inserted

Employer's Requirements

SMM7 (A36) CESMM (A2)

- Accommodation
- Furniture
- Telephone
- Equipment
- Transport
- Attendance

Site Accommodation

SMM7 (A36,41) CESMM (3.1)

- Offices
- Stores
- Canteen/welfare
- Toilets
- Drying and first aid
- Workshops and laboratories
- Foundations and drainage
- Rates and charges
- Erection and fitting out
- Furniture
- Removal
- Transport

Management and Staff

SMM7 (A32,40) CESMM (A3.7)

- Site Manager
- General foreman
- Engineer
- Planning engineer
- Foreman
- Assistant engineer
- Quantity surveyor
- Assistant quantity surveyor
- Clerk/Typist
- Security/watchman

Attendant Labour

SMM7 (A42) CESMM (3.7)

- Unloading and distribution
- Cleaning
- Setting-out assistants
- Drivers and pump attendance
- General attendance
- Scaffold adaptation

Facilities and Services

SMM7 (A34,42) CESMM (A3.2)

- Power/lighting/heating
- Water
- Telephones
- Stationery and postage
- Office equipment
- Computers
- Humidity/temperature control
- Security and safety measures
- Temporary electrics
- Waste skips

Contract Conditions

SMM7 (A20) CESMM (A1)

- Fluctuations
- Insurances
- Bonds
- Warranties
- Special conditions
- Professional fees

Mechanical Plant

SMM7 (A43) CESMM (A3.3)

- Crane and driver
- Hoist
- Dumper
- Forklift
- Tractor and trailer
- Mixer
- Concrete finishing equipment
- Compressor and tools
- Pumps
- Fuel and transport for plant

Non-mechanical Plant

SMM7 (A44) CESMM (A3.6)

- External scaffolding
- Internal scaffolding
- Hoist towers
- Mobile towers
- Small tools and equipment
- Surveying instruments

Temporary Works

SMM7 (A36,44) CESMM (A2,3)

- Access routes
- Hardstandings
- Traffic control
- De-watering
- Hoarding
- Fencing
- Notice board
- Shoring and centring
- Temporary structures
- Protection

Miscellaneous

SMM7 (A33,35)

- Setting out consumables
- Testing and samples
- Winter working
- Quality assurance
- Site limitations
- Protective clothing

Fig 15.1 *Preliminaries checklist*

against measured work. It is becoming more common for these items to be considered as part of the general site overheads because very often these facilities are available to all trades and should be assessed using the tender programme.

A typical sequence of events for pricing preliminaries is:

1. Make notes of general requirements, such as temporary works and sub-contract attendances, when pricing the bill.
2. Prepare a site layout drawing showing the position of accommodation, access routes, storage areas, and services. Inspect site features and check feasibility of proposals during site visit.
3. Use the tender programme for planning staff, plant and temporary works requirements.
4. Read the client's specific requirements and all the tender documents.
5. Price the project overheads sheets.

It is essential that the contractor has standard sheets which give all the main headings for pricing preliminaries. He cannot depend on the descriptions given in the tender documents because they are not necessarily complete. The COEP offers the comprehensive set of forms, but some estimators find there is too much detail for the average job and so a simplified checklist is given in Fig. 15.1. A detailed examination of the items in the checklist is presented in Figs 15.2(a) to 15.2(j).

A detailed programme is an important aid to the accurate pricing of preliminaries because most of the general facilities are related to when the construction activities are carried out. The estimator or planning engineer can superimpose the main elements of the project overheads on the tender programme. The example of a tender programme given earlier shows staff and principal plant durations. Other items drawn from the programme are general plant, scaffolding, fluctuations, attendant labour, temporary works, traffic management and so on. There is an opportunity here for the contractor to be innovative and develop methods which might give a competitive advantage over other tenderers. A typical example is for wall cladding to be fixed by men working from mechanical platforms as opposed to standing scaffolding. This not only reduces the access costs but also cuts the overall contract duration with shorter erection and dismantling periods. Shorter programmes bring about further savings by reducing the staff, overheads and accommodation costs.

(a) Employer's Requirements

Accommodation	Offices, toilets, conference room, stores, laboratories and car parking space may be required depending on client's specific requirements

Furniture	If none stated, assume client providing own furniture

Telephone	Telephone and facsimile equipment can be specified including payment of standing charges (call charges are given as a provisional sum)

Equipment	Technical testing equipment and surveying instruments Protective clothing

Transport	Vehicles for employer's staff or consultants, fuel and maintenance Transport to suppliers to inspect production of components

Attendance	Drivers, chainmen, office cleaners, and laboratory assistants

NOTES :

SMM7 states that, where the employer requires accommodation on site, heating, lighting and maintenance are deemed to be included.

Notice boards are often given as a specific requirement but invariably will be provided by a contractor for information and advertising. (See (e) Temporary Works)

Fig 15.2 *Pricing preliminaries. (a) Employer's requirements. (b) Management and staff. (c) Facilities and services. (d) Mechanical plant. (e) Temporary works. (f) Site accommodation. (g) Attendant labour. (h) Contract conditions. (i) Non-mechanical plant. (j) Miscellaneous.*

(b) Management and Staff

Site Manager	Required on most sites. Calibre of staff depends on size and complexity of project
General Foreman	Day to day management of labour and plant. Coordination of labour—only sub—contractors
Engineer	Analysis of building methods, setting—out and quality control. Services engineer to coordinate specialist services contractors
Planning Engineer	Master programme during mobilization. Up—dating exercises and short—term programmes
Foreman	Consider structure, finishings and snagging. Add non—productive time for trades foremen
Assistant Engineer	Setting out work, external works and internal fabric. Scheduling materials and attendance on sub—contractors
Quantity Surveyor	Some involvement on all jobs, particularly at beginning and end
Assistant quantity surveyor	For large contracts with complex valuations and control of sub—contractors' accounts and bonus payments
Clerk/Typist	General admin duties on site. Checkers. Telephone, post and reception duties
Security/Watchman	Usually employing security services. Important when fittings and furniture arrive

NOTES :

All—in average rates for each category of staff will be provided by senior management in a way which ensures that individual salaries are not identifiable. Employment costs for salaried staff are calculated on an annual basis with additional costs which arise from:

pension scheme (employer's contribution)
annual bonuses
overtime
training levy
car and expenses

The choice of site managers will depend on job size, complexity, duration,
number of operatives and commitments to nominated and domestic
sub—contractors

(c) Facilities and Services

Power/lighting/heating	Check availability of supplies, connection charges, temporary housings, fittings and consumption costs. Alternatively generators and gas bottles
Water	Provision of service to site, pipework from supply, distribution system, and charges. Bowsers required if no piped service available.
Telephones	Consider number of senior staff in deciding on number of lines, switchboard for larger systems and mobile phone for start of job
Stationary and postage	Average cost per week drawn from analysis of previous projects.
Office equipment	Photocopier, facsimile machine and typewriters are commonly required
Computers	Personal computers required for more complex projects. Security can be a problem; consider portable PCs or robust equipment
Humidity and temperature control	Check specific requirements for de−humidifiers, heaters and attendance. Vulnerable materials include seasoned joinery and suspended ceilings
Security and safety measures	Security firm or own labour for watching site at night and at weekends Intruder alarms, traffic control, fire precautions and fire fighting
Temporary electrics	Transformers, distribution system, boards, leads and site lighting. Electrician may be resident on large building schemes
Waste skips	Regular collection of rubbish skips should be allowed Dustbins can be used to promote cleaner sites

NOTES :

Computers are used on site for material records, valuations and cost monitoring. Standard databases are often used to produce drawing registers and lists of instructions. Additional costs which include maintenance contracts, software and consumables can be as much as the value of the computers. The consumables and maintenance costs associated with photocopiers can be costly.

Humidity control is a complex requirement. If dehumidifiers are used the hire charges are high and additional costs include transport charges, electric power, attendance in removing water and daily monitoring of humidity levels. These measures can only be put in place when the building is enclosed and damp air is prevented from entering the building.

Fuels are deemed to be included in items for testing and commissioning mechanical and electrical work. (SMM7 Y51 and Y81). The contractor should liaise with specialists to ensure the fuel costs are included.

(d) Mechanical Plant

Crane and driver	Determine maximum lift in terms of weight, radius and height clearance. Duration and location on site also needed to select a crane (see below)
Hoist	Usually required for multi–storey external access scaffolds. Consider type (goods or passengers), hire, transport, erect, adapt, and dismantle
Dumper	Difficult to dispense with dumpers for moving excavated material, stone, bricks, blocks and mortar. Include transport, hire and fuel
Forklift	Rough terrain forklift is a good all round tool used for unloading, distributing and hoisting palletised materials. May need attachments
Tractor and trailer	Popular for long drainage runs and kerb laying for roadworks
Mixer	Hire, transport and fuel for concrete and mortar mixers, silos and bunkers. Minimum requirement is a mixer for brickwork and drainage
Concrete finishing equipment	Screeding rails and tamping bars. Curing membranes and vibrators
Compressor and tools	Needed for demolitions and alterations, cleaning inside shutters, preparing stop ends in concrete, drilling and vibrating concrete
Pumps	Water pumps, hoses, fuel, transport and attendance (see attendant labour). Concrete pump hire for mobile or static equipment; check quote for extras
Fuel and transport for plant	Static fuel tank or fuel bowser may be required. Assess additional transport costs for plant from yard to site

NOTES :

Quotations should be obtained for long hire or large capacity plant, such as cranes, forklifts and concrete pumps. Consult plant suppliers for advice on running costs and fuel consumption.

Cranes need ancilliary equipment such as slings, concrete skips and lifting beams. Mobile cranes need sufficient space for outriggers, and hire charges are usually from time of leaving depot to return.

A tower crane will incur costs to transport, erect, adapt and dismantle, as well as a foundation or rail tracks, power source, fuel and operator. In some instances tower cranes may not be feasible owing to wind limitations or air space rights which cannot be infringed.

(e) Temporary Works

Access routes	Plot layout on site plan taking advantage of hardcore under roads and buildings. Allow for maintenance and making−up levels on completion
Hardstandings	Additional areas are required for storage, site huts, and lay−down areas for materials such as pipes and reinforcement. Reinstatement.
Traffic control	Check specification and statutory obligations. Use programme and site layout drawing to determine equipment needed and hire periods
De−watering	Establish type of system required. Quotation and advice from specialist needed for a well−point system
Hoarding	Serves to protect the public and forms a secure barrier around the site or contractor's compound. Costs to hire, buy, erect, adapt and dismantle
Fencing	Temporary or permanent fencing to maintain security at perimeter of site, protect trees, mark a boundary and to form site compound
Notice board	Notice board and local signage help drivers and visitors find the site, satisfy curiosity and provide cheap advertising. Check client's requirements
Shoring & Centring	Consider design, duration, and hire or buy calculations for falsework/shoring (temporary shoring will incur making good costs on completion)
Temporary structures	Temporary bridges, temporary roofs, facade supports, ramps, viewing platforms, accommodation gantries. May need specialist design input
Protection	High value components and all finishes need to be considered Decorated areas may require additional coats of paint

NOTES :

In poor soil conditions the loss of hardcore under access roads and hardstandings can be substantial. A ground improvement mat may be used with the approval of the contract administrator.

Excavation below ground water level or works affected by rivers or tidal water is normally identified in the bill of quantities. The contractor is therefore responsible for finding an appropriate method for dealing with the problem. Where a full de−watering system is required by a client, the contractors are normally informed at tender stage.

The cost of protection is often undervalued; particularly in the case of building finishes which are difficult to protect during the commissioning and completion of a project. In a large building measures might include laying sheeting and boarding on floors, re−painting walls and woodwork, and a security system which detects responsibility for damage by allowing entry to finished areas with a written permit.

(f) Site Accommodation

Offices	Mobile offices can be established quickly to high standard
	Sectional sheds require set−up costs, foundations and finishes

Stores	For secure protection of high value materials
	Hire or purchase accommodation, or in the building

Canteen/welfare	Use labour strength from programme and add for sub−contractors
	Add for equipment, cooking facilities, furniture and food subsidies

Toilets	Mobile toilet units − check drainage and services available
	Allow for sundries such as soap, towels and cleaning materials

Drying and first aid	Accommodation with lockers and heaters
	First aid room depending on number of people on site

Foundations and drainage	Concrete or sleepers to support cabins
	Wherever possible connect drainage to live sewer

Rates and charges	Local authority charges are payable on temporary accommodation
	May need to acquire land for site establishment

Erection and fitting out	Materials for erection and fitting out
	Labour may be here or as Attendant Labour

Furniture	Company owned desks, chairs and cabinets usually available
	Replacements can be provided from secondhand market

Removal	Consider the need to re−site the accommodation during the job
	Taking down can be priced as Attendant Labour

Transport	Often priced as transport to site only if follow on work expected
	Cranage off−loading and loading accommodation

NOTES:

The estimator should keep records of hire rates for mobile accommodation and compare with the purchase of sectional buildings. The capital cost of timber buildings can be divided by the life span (plus an allowance for repairs and renovation) to arrive at an equivalent hire rate. There is additional labour in erecting and dismantling timber buildings together with the provision of sundry materials such as sleepers for foundations, felt for roofing, glass, insulation board lining and so on.

Additional hire during the defects liability period may be required for a foreman's office and storage container. Minimum requirements for accommodation and health and welfare are given in the Construction Regulations.

(g) Attendant Labour

Unloading and distribution	Envisage an unloading and handling labour gang; adding drivers for dumper and forklift. Balance requirements with attendant labour in bill rates
Cleaning	Daily and weekly tidying−up and cleaning may be carried out by the general gang or casual labour
Setting−out assistants	Each setting−out engineer will need some assistance from a chainman.
Drivers and pump attendance	Each item of mechanical plant will need some operator time; for example: a water pump might require 1.5 man hrs/day
General attendance	Clearing away rubbish for sub−contractors, adapting scaffolding and materials distribution (may be priced elsewhere)
Scaffold adaptation	An assessment of time is required and rates should be obtained from the scaffolding specialist

NOTES :

A site which relies on scaffold hoists will need more labour than that which has a tower crane. Where multi−storey buildings have blockwork walls internally, a considerable amount of labour for distribution is needed.

It can be argued that general attendance for sub−contractors is already included in the tender by assessing all the site facilities in the preliminaries. Special attendance can also be dealt with in this section, although some estimators prefer to price specific attendances for nominated sub−contractors in the bill of quantities, which is presumably what is envisaged in the standard method of measurement.

(h) Contract Conditions

Fluctuations *	Calculation of fluctuation costs includes changes forecast for preliminaries See fluctuations chapter for factors to be considered

Insurances	A quotation (or guide rate) can be obtained for all−risks insurance. The all−in rate for labour usually includes the Employer's Liability insurance

Bonds	Cost of bond is based on contract value and duration inc defects liability period. Check that bond is acceptable and bank facilities are not exceeded

Warranties	More common with contractor−designed elements which are usually backed up with warranties from specialists and designers

Special conditions	Check changes to standard conditions, poor documentation and that form of contract, specifications and SMM's are up−to−date

Professional fees	Alternative bids which include a design element, tests on materials, QS services, and legal fees to check contractual arrangements. Land surveys

NOTES :

*   Most fluctuations in prices will relate to production costs, not preliminaries. It is therefore preferable to calculate fluctuations independently and transfer the result to the tender summary.

Not all losses can be recovered under an insurance policy. There could be many losses on a site which each fall below the policy excess agreed with the insurers. An estimate of an average value of losses should be added to the tender.

Where applicable, a Parent Company Guarantee would provide a measure of protection, at little cost to the client.

Tendering costs are not normally added to an individual tender. They become part of the general company overhead.

197

```
┌─────────────────────────────────────────────────────────────────┐
│                                                                   │
│  (i) Non–mechanical Plant                                         │
│                                                                   │
└─────────────────────────────────────────────────────────────────┘
```

| External scaffolding | Quotation needed for erection and hire of scaffold for larger schemes |
| | Programme used to assess duration on each section of work |

| Internal scaffolding | Birdcage scaffold for large voids, decking for ceilings and ductwork |
| | Consider lift shafts, stair wells, high walls & inner skin of external walls |

| Hoist towers | Extra cost to provide enclosure for mechanical hoist |
| | Platforms to unload materials at each level |

| Mobile towers | For short duration and localised work |
| | Consider also platform hoists or scissor lifts (mechanical plant) |

| Small tools and equipment | Hand tools such as drills, power saws, picks and shovels, rubbish chutes, |
| | bending machines, bolts crops and traffic cones |

| Surveying instruments | Purchase, hire or internal charges, maintenance and consumables |
| | Check requirements of engineering staff listed in supervision section |

NOTES :

For scaffolding the following should be considered:

a. Hire charges   g. Baseplate support
b. Transport costs   h. Debris netting
c. Labour costs   j. Polythene sheeting
d. Losses   k. Temporary roofing
e. Adaptation   l. Platforms to land materials
f. Safety measures

The loads imposed on scaffolding should be considered, in particular the use of scaffolding for short–term storage of block stone or bricks.

There is a relationship between the amount of labour in a job and the small tools and equipment costs . Some companies add a percentage to the all–in labour rate for small tools others include a sum in the preliminaries based on a percentage of total labour costs for the job. Clearly feedback from previous contracts is needed so that the estimator has a realistic guide for this item.

(j) Miscellaneous

Setting out consumables	Pegs and profile boards, tapes and refills Larger projects need concrete and steelwork for site stations
Testing and samples	Concrete cube testing is calculated from the specified frequency of tests Samples of materials may be supplied free but composite panels have a cost
Winter working	Additional protection for operatives, heating, and site lighting Reduced productivity, location of work, heated concrete etc
Quality assurance	For large schemes, a significant proportion of the supervisor's time may be dedicated to the control and monitoring systems. See Management & Staff.
Site limitations	Check employer's requirements such as access restrictions, control of noise, weight of vehicles, protection of services, etc. See Temporary Works.
Protective clothing	Staff and directly employed people will need protective clothing, depending on time of year. Safety hats/clothing for employees and labour—only s/c's

NOTES :

The responsibility for quality assurance remains with all the parties and everyone in the organization. The establishment of a quality system will produce quality statements, a set of company procedures, training, and control mechanisms. The cost is usually carried by the general off–site overhead. It is often argued that the cost of setting up and implementing a quality system is off–set by the benefits which result.

Pricing the preliminaries bill

Both SMM7 and CESMM3 recommend that 'fixed' and 'time-related' charges are identified separately in a bill of quantities. SMM7 defines them as follows:

1. A fixed charge is for work the cost of which is to be considered as independent of duration.
2. A time-related charge is for work the cost of which is to be considered as dependent on duration.

There are certain items which are difficult to allocate such as the use of specialist plant. A crane, for example, may be on site for two weeks; should this be classed as a fixed or time-related charge? For many schemes, all general plant and facilities are divided by the duration to produce equal sums for monthly payments. Similarly, the staff costs are difficult to share between valuations because they are usually higher at the beginning of a contract and taper off gradually towards the end.

When the estimate has been adjudicated by management the estimator will allocate sums of money to the preliminaries bill. This is an opportunity to ensure that a satisfactory (and possibly positive) cashflow position is secured. In particular, the contractor needs to identify setting-up costs which should be claimed in the first valuation as fixed charges. If all site and general overheads were reimbursed in proportion to time the contractor would have more expenditure than income and this poor cashflow position would persist during most of the contract duration. The early fixed charges often include:

Employer's requirements
Accommodation
Furniture
Install telephone
Provide equipment
Transport charges

Supervision
Hotel expenses
Planning
Procurement

Services
Installation charges
Office equipment

Mechanical plant
Transport of plant
Purchase of plant

Temporary works
Design and purchase of structures
Access routes and hardstandings
Enclosures
Dewatering, piling and formwork

Site accommodation
Transport and cranage
Purchase costs
Foundations and furniture
Erection and fitting out

Contract conditions
Insurance premium
Bond
Professional fees
Initial land/building surveys

Non-mechanical plant
Scaffold erection
Small tools and equipment

Miscellaneous
Setting-out consumables
Samples
Quality planning
Protective clothing

The costs of dismantling site facilities and cleaning are much smaller by comparison and are rarely priced as separate fixed charges.

16 *Cashflow forecasts*

Introduction

At tender stage, a contractor sets up his financial and time objectives by calculating construction costs and producing a project programme. By linking the two sets of data, an estimator can first, help a client produce his forecast of payments and second, compare this with his likely payments (to suppliers and sub-contractors) to produce his own cashflow forecast. In this way a contractor is in a unique position to give accurate information to the building team.

There is seldom enough time at tender stage to produce detailed cashflow forecasts. The contractor knows that the objective is to win the contract, so there must be good reasons for putting in this effort. Obviously if a client has asked for a tender-stage programme and cashflow forecast it must be done. The contractor may also need to assess the cashflow benefit of taking on a job, because this is part of his assessment of risk. A spreadsheet model may be able to answer questions such as: are there any sudden cash commitments? How much early money will be needed to make the contract self-financing?

Cashflow calculations

There are two methods commonly used to predict the value of project work over time. First, cost models can be used at various pre-contract stages to produce approximate forecasts and second, the estimator's calculations form the basis of a more detailed technique.

If a client needs a schedule of payments, the simplest model would be a straight-line relationship of value against time from which the client's commitments can be shown (see Fig. 16.1). The assumption is that all payments are of an equal amount based on the full value divided by the number of payments with a small adjustment for retention money.

A slightly more sophisticated technique assumes that a construction project

week nos.	gross value £1000's	retention 5% £1000's	interim payments £1000's
0	0	0	
1	22	1	
2	44	2	
3	65	3	
4	87	4	
5	109	5	
6	131	7	83
7	152	8	
8	174	9	
9	196	10	
10	218	11	165
11	239	12	
12	261	13	
13	283	14	
14	305	15	
15	327	16	269
16	348	17	
17	370	19	
18	392	20	
19	414	21	352
20	435	22	
21	457	23	
22	479	24	
23	501	25	
24	522	26	455
25	544	27	
26	566	28	
27	588	29	
28	609	30	538
29	631	32	
30	653	16	
31			
32			637

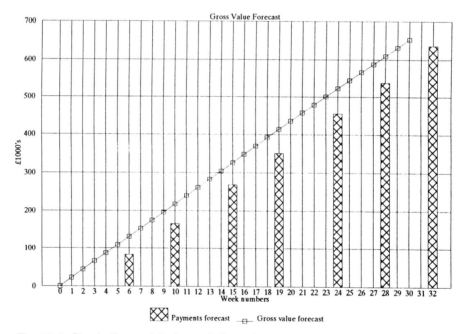

Fig 16.1 *Simple linear plot of cumulative value*

accumulates costs in a way which can be represented by an S-curve graph (Fig. 16.2). This model is based on the presumption that only one quarter of the costs are incurred during the first third of a project duration, half the costs arise from the second third (in a linear fashion) and the remaining quarter of costs occur in the final third of the project duration. The S-curve can be created manually by drawing a straight line between the first and second 'third' points and sketching the parabolic end portions. The contractor's cumulative cost curve can be superimposed on the chart by deducting the profit margin from the cumulative value curve. The S-curve method is, of course, a theoretical technique which is difficult to change to take account of the nature of individual projects and contractors' pricing methods, but is successfully used at early stages when detailed pricing information is not available.

The GC/Works 1/Edition 3 form of contract has introduced the S-curve principle as a basis for stage payments. The printed form gives charts for projects with contract values over £5.5 million, and others are available for smaller jobs. Figure 16.3 shows the S-curve produced from the data given for a 100-week project. Clearly, the Project Manager is able to predict the client's payments and a great deal of time is saved each month in producing valuations. The same is true for the contractor, but there could be a shortfall in payments if the establishment costs are high.

An estimator's calculations and programme are the best starting-point for a contractor's cashflow forecast. The rates calculated by the estimator can be linked with the relevant activities on a tender programme. The total costs associated with each activity are divided by the duration to arrive at a weekly cost. The information used to produce the value curve is simply taken from the sums inserted in the client's bill with adjustments for retention. The contractor's income can be predicted by taking the value at each valuation date and allowing a delay for payments.

To calculate the weekly cost commitment (the contractor's outgoings) each element of cost should be viewed separately. This is because spending on labour, materials, plant and specialist contractors develops in different ways. Direct labour, for example, is a weekly commitment, and credit arrangements for materials can delay payments for up to nine weeks. Expenditure and income can be plotted on a graph against time. The combined effect is a cumulative cashflow diagram which will show the extent to which the job needs financing by the contractor.

week nos	gross value £1000's	cost Forecast £1000's	
0	0	0	
1			
2			
3			
4			
5			
6			
7			
8			
9			
10	163.25	148.75	< quarter value at third duration
11			
12			
13			
14			
15			
16			
17			
18			
19			
20	489.75	446.25	< half value in second third duration
21			
22			
23			
24			
25			
26			
27			
28			
29			
30	653	605	

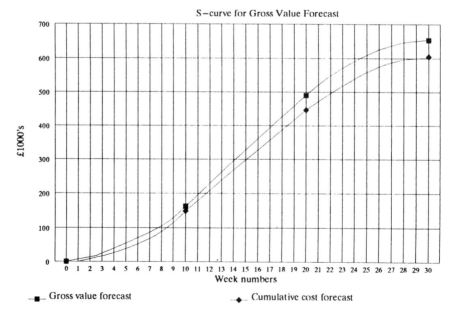

S−curve for Gross Value Forecast

£1000's

Week numbers

—■— Gross value forecast   —♦— Cumulative cost forecast

Fig 16.2 *Simple S-curve for cumulative value calculated at 'third' points*

week nos	%Payable before retention
0	0
5	0.85
10	2.85
15	5.89
20	9.84
25	14.58
30	19.99
35	25.97
40	32.37
45	39.10
50	46.02
55	53.03
60	59.99
65	66.80
70	73.33
75	79.46
80	85.08
85	90.06
90	94.28
95	97.64
100	100

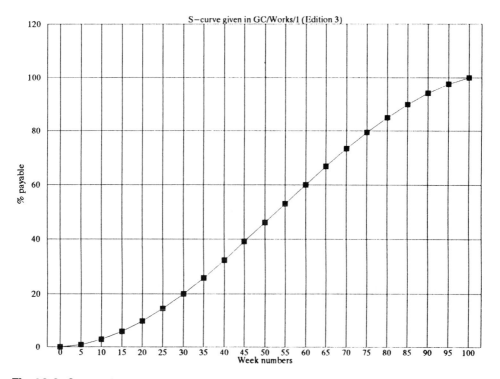

Fig 16.3 *S-curve based on GC/Works/1 Edition 3 data*

Example of a contractor's cashflow forecast

The following analysis was carried out by an estimator who was successful in winning a contract for Fast Transport Ltd. He received the enquiry, in the form of drawings and a bill of quantities, from a local QS practice. Interim payments are to be made monthly and retention has been set at 5%. A sum of £18 000 has been included for profit, and the rates in the bill of quantities exclude any on- or off-site overheads. Following a careful review of the estimate, the construction director asked for all sub-contract discounts to be taken from the estimate but made an adjustment to the profit margin.

During the mobilization period the estimator was asked to produce a forecast of payments for the client and a cashflow forecast for the commercial department. The estimator priced the bill of quantities analytically using estimating software. The first task was to ensure that all the adjustments made at the adjudication stage had been made to the rates in the bill of allowances. This fully adjusted bill would become the budget for the construction team to monitor the financial progress of the job.

The next stage was to assign sums of money to each activity on the tender programme. This was done using a spreadsheet program because each cell on the screen can hold text, graphics, numbers or formulae. Each row on the spreadsheet can show weekly values with shading used to locate the activity bars. Figure 16.4 shows the tender programme with contract values taken from the bill of quantities which was submitted to the client. The main assumptions used for this graph were that:

1. Each activity will be completed on time.
2. Sums are divided equally for the duration of an activity.
3. Provisional sums and daywork are not included.

The total for preliminaries has been split equally over the whole duration. This is not a good interpretation for the contractor because he will incur more expenditure setting up his site facilities at the start of a job. The final presentation to the client did not include the costed programme or a graph. The commercial department felt that the client would want a simple list of payments, and included one in a letter to the quantity surveyor.

The contractor's cashflow diagram was created on a spreadsheet program which had the facility to use multiple sheets in the same file. Figure 16.5 shows the contents of the costed programmes for labour, plant, materials and sub-contractors. The bottom sheet was used to consolidate the cost commitment drawn from each of the other sheets with the forecast income. The cashflow forecast is simply the difference between income and expenditure. The results

Activity	Contract Value £	Sept 1	2	3	4	Oct 5	6	7	8	Nov 9	10	11	12	13	Dec 14	15	16	17	18	Jan 19	20	21	22	23	Feb 24	25	26	27	Mar 28	29	30	Apr 31	32
1 Mobilization & set up	6450	3225	3225																															
2 Excavation and filling	21480		7160	7160	7160																													
3 Foundations formwork	19260			3852	3852	3852	3852	3852																										
4 Foundations concrete	27800				6950	6950	6950	6950																										
5 Underslab drainage	5210					2605	2605																											
6 Concrete ground floors	16470							8235	8235																									
7 Columns formwork	6007					1502	1502	1502	1502																									
8 Columns concrete	2700						1350		1350																									
9 Floors and beams formwork	22220							2778	2778	2778	2778	2778	2778	2778	2778																			
10 Floors and beams concrete	21140								7047	7047			7047		7047																			
11 External walls	78580										9823	9823	9823	9823	9823	9823	9823	9823															
12 Roof timbers	9350																3117	3117	3117															
13 Roof covering	19030																	9515	9515															
14 Windows	25500															6375	6375	6375	6375															
15 Services 1st fix	43000														7167	7167	7167	7167	7167	7167														
16 Plasterwork and partitions	43500																		5438	5438	5438	5438	5438	5438	5438	5438								
17 Joinery	41670																			5209	5209	5209	5209	5209	5209	5209	5209	5209						
18 Ceilings	23000																					4800	4800	4800	4800	4800								
19 Services 2nd fix	28000																								5600	5600	5600	5600	5600					
20 Painting	12300																						2050	2050	2050	2050	2050	2050	2050					
21 Floor coverings	12300																													4100	4100	4100		
22 External work and drainage	60920		6092	6092																			6092	6092	6092	6092	6092	6092	6092	6092	6092			
23 Preliminaries	107450	3582	3582	3582	3582	3582	3582	3582	3582	3582	3582	3582	3582	3582	3582	3582	3582	3582	3582	3582	3582	3582	3582	3582	3582	3582	3582	3582	3582	3582	3582		
Gross weekly value forecast	653.3 k	6.81	20.1	20.7	21.5	18.5	19.8	28.9	9.21	21.6	16.2	16.2	23.2	16.2	23.2	20.8	30.1	30.1	35.2	32.1	21.4	14.2	22.4	22.4	24.9	30.5	27.1	27.1	26	21.4	7.88		
Gross interim valuation	653.3 k					69.1			74.4				93.4				104				125							105			82.26		
Net interim payments	620.7 k					65.6			70.7				88.7				98.7				119							98.7			78.15		
After completion	32.7 k																															32.67	
week number		1	2	3	4	5	6	7	8	9	10	11	12	13	14	15	16	17	18	19	20	21	22	23	24	25	26	27	28	29	30	31	>>

Fig 16.4 *Client's cashflow forecast produced by the estimator*

Activity	Value forecast for Labour					1	2	3	4	5	6	7	8	9	10	11	12
	lab	plt	mat	sub	tot												
1 Setting up and setting out	2800	2400	1250		6450	1400	1400										
2 Excavation and filling	7800	7980	5700		21480		2600	2600	2600								
3 Foundations formwork	12480		6780		19260			2496	2496	2496	2496	2496					
4 Foundations concrete	4600		23200		27800				1150	1150	1150	1150					
5 Underslab drainage	1920	1750	1540		5210					960	960						
6 Concrete ground floors	3420	1750	11300		16470							1710		1710			
7 Columns formwork	3907		2100		6007					977	977	977	977				

Activity	Value forecast for Plant					1	2	3	4	5	6	7	8	9	10	11	12	13
	lab	plt	mat	sub	tot													
1 Setting up and setting out	2800	2400	1250		6450	1200	1200											
2 Excavation and filling	7800	7980	5700		21480		2660	2660	2660									
3 Foundations formwork	12480		6780		19260													
4 Foundations concrete	4600		23200		27800													
5 Underslab drainage	1920	1750	1540		5210					875	875							
6 Concrete ground floors	3420	1750	11300		16470							875		875				
7 Columns formwork	3907		2100		6007													

Activity	Value forecast for Materials					1	2	3	4	5	6	7	8	9	10	11	12	13	14
	lab	plt	mat	sub	tot														
1 Setting up and setting out	2800	2400	1250		6450	825	825												
2 Excavation and filling	7800	7980	5700		21480		1900	1900	1900										
3 Foundations formwork	12480		6780		19260			1356	1356	1356	1356	1356							
4 Foundations concrete	4600		23200		27800				5800	5800	5800	5800							
5 Underslab drainage	1920	1750	1540		5210					770	770								
6 Concrete ground floors	3420	1750	11300		16470							5650		5650					
7 Columns formwork	3907		2100		6007					525	525	525	525						

Activity	Value forecast for sub-contractors					1	2	3	4	5	6	7	8	9	10	11	12	13	14	15
	lab	plt	mat	sub	tot															
1 Setting up and setting out	2800	2400	1250		6450															
2 Excavation and filling	7800	7980	5700		21480															
3 Foundations formwork	12480		6780		19260															
4 Foundations concrete	4600		23200		27800															
5 Underslab drainage	1920	1750	1540		5210															
6 Concrete ground floors	3420	1750	11300		16470															
7 Columns formwork	3907		2100		6007															

			1	2	3	4	5	6	7	8	9	10	11	12	13	14	15	16
Cumulative labour value	delayed	1 weeks	2.7	9.7	17.8	25.3	32.2	39.5	49.0	53.5	59.6	67.3	75.0	84.0	91.7	100.7	106.6	
Cumulative plant value	delayed	4 weeks				2.4	6.7	13.8	17.6	19.7	21.9	23.9	25.2	27.9	29.1	30.2	32.0	
Cumulative materials value	delayed	6 weeks					1.0	5.9	11.5	20.9	29.8	39.4	54.1	56.8	69.0	75.6		
Cumulative sub-contract value	(week after payment received)						2.8	2.8	2.8	2.8	3.1	3.1	3.1	3.1	3.1	3.6		
Weekly cost commitment			2.7	9.7	17.8	27.7	40.9	57.1	75.2	87.4	105.1	124.0	142.7	169.0	180.6	203.0	217.7	
			1	2	3	4	5	6	7	8	9	10	11	12	13	14	15	16

Fig 16.5 *Multiple sheets for cashflow analysis*

were plotted, by the program on a single diagram (Fig. 16.6) and submitted to the commercial manager in the form of a graph linked to the consolidated costed programme.

Since all the information was held in a single spreadsheet file, the computer was able to answer 'what if' questions which allowed the commercial manager to minimize the borrowing requirement. Clearly this information can be the basis of a cost-monitoring system (which is beyond the scope of this book). The cumulative value curve can help the construction team in monitoring progress of their work and that of the sub-contractors. This technique is sometimes adopted by management contractors as a check on the progress of their works contractors.

Proposed Offices for Fast Transport Limited

Tender programme No: T354/P1

Contract value and time-phased programme

Activity	lab	plt	mat	sub	tot
1 Setting up and setting out	2800	2400	1250		6450
2 Excavation and filling	7800	7980	5700		21480
3 Foundations formwork	12480		6780		19260
4 Foundations concrete	4600		23200		27800
5 Underslab drainage	1920	1750	1540		5210
6 Concrete ground floors	3420	1750	11300		16470
7 Columns formwork	3907		2100		6007
8 Columns concrete	750	250	1700		2700
9 Floors and beams formwork	14570		7650		22220
10 Floors and beams concrete	3750	1750	15640		21140
11 External walls	36700		41880		78580
12 Roof timbers	4140		5210		9350
13 Roof covering	1350		1280	16400	19030
14 Windows				25500	25500
15 Services 1st fix				43000	43000
16 Plasterwork and partitions				43500	43500
17 Joinery	2360	5310		34000	41670
18 Ceilings	1800		1400	19800	23000
19 Services 2nd fix				26000	26000
20 Painting				12300	12300
21 Floor coverings				12300	12300
22 External work and drainage	16700	12550	19870	11800	60920
Total for measured work	119047	28430	151810	246600	545887
PRELIMINARIES	57430	35490	11290	3300	107450
Gross weekly value forecast					653.34

Weekly programme (1994–1995)

Months across the programme: September (weeks 1–3), October (4–7), November (8–13), December (14–16), January 1995 (17–20), February (21–24), March (25–30), April (31–35).

Activity	phased weekly values
1 Setting up and setting out	w1 3225, w2 3225
2 Excavation and filling	w2 7160, w3 7160, w4 7160
3 Foundations formwork	w4 3852, w5 3852, w6 3852, w7 3852
4 Foundations concrete	w4 6950, w5 6950, w6 6950, w7 6950
5 Underslab drainage	w5 2605, w6 2605
6 Concrete ground floors	w7 8235, w8 8235
7 Columns formwork	w5 1502, w6 1502, w7 1502, w8 1502
8 Columns concrete	w6 1350, w7 1350, w8 1350
9 Floors and beams formwork	w8–w14 2778 each
10 Floors and beams concrete	w13 7047, w14 7047
11 External walls	w9–w16 9823 each
12 Roof timbers	w15 3117, w16 3117, w17 3117
13 Roof covering	w16 8515, w17 9515
14 Windows	w16–w19 6375 each
15 Services 1st fix	w14–w19 7167 each
16 Plasterwork and partitions	w19–w26 5438 each
17 Joinery	w21–w28 5209 each
18 Ceilings	w25–w29 4600 each
19 Services 2nd fix	w26–w30 5600 each
20 Painting	w22–w27 2050 each
21 Floor coverings	w28 4100, w29 4100, w30 4100
22 External work and drainage	w2 6092, w3 6092, w21–w30 6092 each

Forecast summary (£1000's)

Row	weekly / monthly values (weeks 1→35)
PRELIMINARIES	3.6 each week
Gross weekly value forecast	3.2, 16.5, 17.1, 18.0, 14.9, 16.3, 23.3, 5.6, 18.1, 12.6, 12.6, 19.6, 12.6, 19.6, 17.0, 26.5, 26.5, 31.6, 28.5, 17.8, 10.6, 18.8, 21.3, 18.8, 26.9, 23.6, 23.6, 22.4, 17.8, 4.1
Gross weekly value forecast (cumulative)	6.6, 25.9, 47.6, 69.1, 86, 107, 134, 166, 165, 181, 198, 221, 237, 260, 261, 311, 341, 376, 408, 430, 444, 466, 489, 513, 544, 571, 598, 624, 646, 653
Net monthly income (with 5% retention)	2.7, 9.7, 27.7, 41, 57, 75, 87, 105, 124, 143, 169, 181, 203, 216, 233, 254, 265, 313, 321, 331, 343, 350, 446, 453, 460, 469, 540, 543, 549, 543, 543, 562, 623, 626, 629, 635
Gross cost commitment forecast	6092, 6092, 3117, 3117, 9515, 9515, 6375, 6375, 6375, 7167, 7167, 7167, 7167, 5209, 5209, 5209, 5209, 5209, 5209, 4600, 4600, 4600, 4600, 4100, 4100, 4100, 6092, 6092, 6092, 6092, 521, 582, 628, 637, 653
Cash flow forecast (£1000's)	-3, -10, -18, 4, 5, 6, 7, 8, -10, 49, 31, 12, -6, -33, 22, 7, -8, 59, 11, 3, -7, 93, 93, -3, 83, 74, 2, -8, 18
Week no.	1, 2, 3, 4, 5, 6, 7, 8, 9, 10, 11, 12, 13, 14, 15, 16, 17, 18, 19, 20, 21, 22, 23, 24, 25, 26, 27, 28, 29, 30, 31, 32, 33, 34, 35

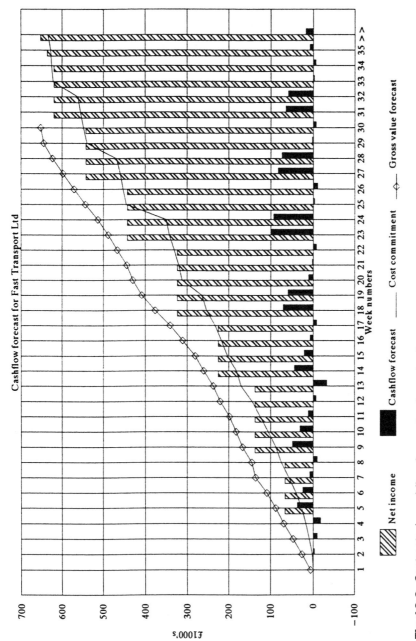

Fig 16.6 *Contractor's cashflow forecast using priced programme*

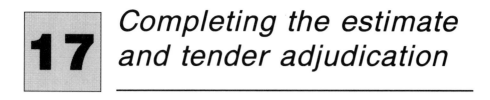

Completing the estimate and tender adjudication

17

Converting an estimate into a tender

Completing the estimate

The estimator must keep clear pricing information so that all the build-ups, assumptions and underlying decisions can be seen and understood by the adjudication team, and the construction staff if the tender is successful. The bills should be extended and totalled, with separate subtotals being produced for the four basic elements of labour, plant, materials and sub-contractors. With computer methods, rates will be put into the database as the estimator attaches resources to the bill items.

All estimating methods (manual or computer) must include reliable techniques to ensure that the analysis of the estimate presented to management is correct. Mathematical checks are relatively simple and can be carried out by estimating clerks or computers. It is perhaps more important to develop a method which can identify a mistake in resourcing (particularly the main items), collecting to summaries and input of quantities if computers are used. Other pricing checks include:

1. Rates must apply to the unit of measurement (for example, an item measured in m^2 is not priced as m^3).
2. All items on each page have been priced or included elsewhere.
3. Major items needing costly resources should be reconciled with resource summary sheets.

Table D of the estimator's report, given in the COEP, shows how the resources for each trade can be identified from the priced bill of quantities, and compared with summary sheets for each of the resources. For example: the materials column might show £15 500.00 for materials in the brickwork section of the bill. A materials summary would include the information received from suppliers such as the cost of bricks, mortar, ties and insulation. The estimator would expect the material resource sheet to give a figure slightly less than £15 500 because of sundry items which would be included in the bill but missing from the materials summary.

The total in Table D for sub-contractors should be the same as the summary used to list the sub-contractors used in the tender. The resource summary sheets provide vital information for three reasons:

1. Reconciliation and checking of bill totals for labour, plant, materials and sub-contractors.
2. They give management a breakdown of resources so they know where to focus their efforts at the adjudication stage.
3. Adjustments to resource costs can be calculated on the sheets before being taken to the tender summary. (This could be done quickly using a computer estimating package but some people prefer to have the full picture on paper before changing the estimate.)

The tender programme must be completed so the estimator can examine the resources estimated in time on the programme with those which he has expressed in cash terms in the estimate. A common problem is the need to maintain excavators on site during groundwork operations. The planner may believe that an excavator is needed on site for ten weeks whereas the estimator has allowed enough money for only six weeks. This reconciliation needs to be carried out jointly by the estimator and planner, and clearly presented to management for final review.

Estimator's report

Once the bill of quantities has been priced and checked, the summaries and reports can be filled in. Each company has its own standard layout for these forms and a comprehensive example is given in the COEP. It would be wrong to make any changes after this stage to the bill, whether produced manually or by computer, because the analysis sheets and summaries would also have to be changed. As long as adjustments are carefully recorded, they can be built back into the bill of quantities later; in particular when the inked-in copy of the bill is called for. This is one reason why priced bills of quantities for building projects are not submitted with the tender.

The estimator's report must show a full breakdown of costs for management. Decisions can then be reached and the summary of the tender can be compiled. The Estimate Analysis form (Fig. 17.1) and Tender Summary form (Fig. 17.2) will be satisfactory for all sizes of contracts. The difference between this and the final summary given in the COEP is that here the Preliminaries are broken down into resource elements. Each item has a reference number for easy identification within the text. If an estimate report has been produced by a computer system, there should still be an independent check of major resources using the summary sheets.

Comments on the Estimate Analysis form

The Estimate Analysis form shown in Fig. 17.1 is an intermediate step towards filling in the Tender Summary form (Fig. 17.2). If a computer estimating system is being used, most of this information will be available in printout form.

Item 1. For a manual system, transfer the total of all measured work from the extended priced bills, include provisional sums and dayworks and take special care to include any provisional sums given in the preliminaries bill.

Items 2-6. From the relevant resource sheets enter the totals incorporated into the measured bills under their respective headings. PC sums drawn from the bill of quantities include discounts at this stage but will be reduced to net values on the Tender summary form. The value of the contractor's own measured work is then found by deducting the sub-contracts and PC and provisional sums from the bill total. The breakdown of measured work can be found in four ways:

217

CB Construction Limited	Project:	Fast Transport Ltd
ESTIMATE ANALYSIS	Date:	Jul 1994
	Tender no:	T354

	Bill totals	Estimator's Adjustments	Estimate
1 Bill total	664705	−56288	608417
DEDUCT			
2 Dayworks	14600		14600
3 Provisional sums	30500		30500
4 Domestic sub−contractors	268535	−29135	239400
5 Nominated sub−contractors	20000		20000
6 Nominated suppliers	3800		3800
7 Own measured work	327270	−27153	300117
8 Breakdown : Lab	136960	−21333	115627
9 Plt	31330		31330
10 Mat	158980	−5820	153160
LABOUR STRENGTH			
11 Contract period stated/offered	30 wks		
12 Labour total	£115,627.00		
13 Average weekly earnings £7.5 40	£300.00		
14 Man weeks	385.4 nr		
15 Average labour strength	12.8 men		
CASH DISCOUNTS			
16 Nominated sub−contractors	500		500
17 Nominated suppliers	190		190
18 Domestic sub−contractors	6713	−728	5985
19 Materials	7949	−291	7658
20 total	15352	−1019	14333

Fig 17.1 *Estimate analysis form*

CB Construction Limited	Project:	Fast Transport Ltd
TENDER SUMMARY	Date:	Jul 1994
	Tender no:	T354

			Estimate	Adjudication Changes	Tender	Analysis
1 Own measured work	Lab		115627		115627	18.2 % of 11
	Plt		31330	−5300	26030	4.1 % of 11
	Mat	nett	145502	−2411	143091	22.5 % of 11
2 Domestic sub−contractors		nett	233415	−12480	220935	34.8 % of 11
3 Nominated sub−contractors		nett	19500		19500	3.1 % of 11
4 Nominated suppliers		nett	3610		3610	0.6 % of 11
	total		548984	−20191	528793	83.2 % of 11
5 Provisional sums	Defined					
6 Preliminaries	Lab		12960		12960	2.0 % of 11
	Plt		36090	1800	37890	6.0 % of 11
	Mat		5780		5780	0.9 % of 11
	Sub		3300		3300	0.5 % of 11
	Staff		32219	5400	37619	5.9 % of 11
7 Price fluctuations	Lab	*	620		620	0.5 % of 1
	Plt	*				
	Mat	*				
	Sub	*	3160	−3160	0	
	Staff	*	1675		1675	4.5 % of staff
8 Water charges	%	*	650		650	0.1 % of 11
9 Insurances	%	*		5000	5000	0.8 % of 11
10 Bond	%	*	1050		1050	0.17 % of 11
11	total		646488	−11151	635337	
12 Overheads			10667	−2691	7976	1.1 % of 17
13 Profit			12000	−1976	10024	1.4 % of 17
14 Dayworks			14600		14600	
15 Provisional sums	Undefined		30500		30500	
	total		714255	−15818	698437	
16 Professional fees	%	*	—		—	
17	TENDER £		714255		698437	

* may be priced in prelims

Fig 17.2 *Tender summary form*

1. By applying an approximate ratio such as 40% labour, 10% plant and 50% materials.
2. By keeping accurate records of the resources used in pricing the works.
3. By analytical pricing and extending each part of the rates to a total.
4. By using the computer printout of resource totals.

It is no longer safe to examine resources using ratios or percentages because so much work today may be carried out by sub-contractors. It is not unusual for an estimator to price the earthwork himself on one contract and use a sub-contractor's quotation for another similar scheme. The Estimator's adjustments are shown in the middle column so that if late prices or mistakes are found they can be recorded before the adjudication meeting.

Item 11. The estimator must clearly state the duration used in calculating preliminaries and say whether this is the period given in the enquiry document or one which the tender programme has yielded.

Items 12-14. Dividing the labour value by the average earnings will give the number of man weeks included in the estimate.

Item 15. The average labour strength is calculated by dividing the number of man weeks, arrived at in 14, by the contract period. The peak labour strengths can be found by reference to the programme, because the size of the workforce on site may well be double the average figure and this does not allow for people employed by the domestic and nominated sub-contractors. An attempt should be made assess the size of the labour force to calculate certain general expenses such as welfare facilities, transport, and protective clothing.

Items 16-19. All discounts offered by suppliers and sub-contractors should be identified because the net value of the resources must be calculated. For nominated suppliers and sub-contractors, refer to the contract conditions for the discounts included in the prime cost sums. JCT contracts have in the past allowed 2.5% on nominated sub-contractors and 5% on nominated suppliers sums but this could change. GC/Works 1/Edition 3 has now dispensed with the discounts, which has set an unwelcome trap for the unwary estimator.

The discounts for domestic sub-contractors and material suppliers will be calculated on the summaries of quotations used. Not all firms are prepared to offer discounts because the benefit of prompt payment is often ignored by main contractors.

Each time an adjustment is made, care must be taken to make a corresponding change to the discount. In the example shown in Fig. 17.1 the estimator has reduced the value of domestic sub-contractors (item 4) by £29 135 and reduced the total discount from £6713 to £5985 in item 18.

Comments on the Tender Summary form

Before the adjudication meeting the estimator will complete the left-hand column of the Tender Summary which gives management an outline of the whole job. In this example the estimator has perhaps gone further than necessary and suggested a total margin of £22 667 comprising £10 667 for overheads and £12 000 for profit. In the event, the margin was reduced to £18 000. The 'adjudication changes' column shows the financial effect of the decisions made by management during the meeting. The most significant changes are to the contractor's own work and domestic sub-contractors, where potential savings have been identified. On the other hand, senior managers are also prepared to add to the estimate and in this case have increased staff costs and insurances. These last two points may have been the subject of a company policy change which the estimator would not have known. Clearly an aggressive stance has been taken with sub-contractors, because the inflation allowance made by the estimator has been removed.

Items 1-4 are taken from the Estimate Analysis and preliminaries (item 6) comes from the summary of project overheads.

Item 5. Provisional sums must now be split between those which are defined (contractor to allow for programming, planning and preliminaries) and those which are undefined (the contractor will be able to recover overhead costs).

Item 7. Fluctuations were dealt with in Chapter 13. This example highlights the treatment of staff costs which are the most difficult to control. Price bargaining may produce savings with suppliers and sub-contractors but seldom works with supervisory staff.

Water charges, insurances and bonds (items 8–10) may have been priced in the site overheads schedule, but can be more accurately calculated when the full estimate is known. Each of these items is governed by the contract sum.

Item 8. Water charges are made by the water company for water consumed during the construction period. The estimator should contact the local water company to establish the charges. These may be a straightforward percentage or scale of charges based on the contract value. Alternatively, the water may be metered.

Item 9. The employer's liability insurance is rarely included today in the average labour rate. This is because labour can be drawn from several sources and a global calculation is carried out in the project overheads, or at the stage when all the parts of the estimate are known. The contract conditions should be examined carefully and changes to the standard conditions noted. Since insurance provisions are notoriously difficult to interpret, the estimator will ask for help from his insurance adviser; and in some cases obtain a quotation before completing the estimate. There are often many small losses suffered by the contractor which are not recovered by an insurance policy. The estimator should

221

ask for records of what the average shortfall may be for the work envisaged. In the example, management decided to add £5000, which was the excess given in the all-risks policy.

Item 10. Performance bonds and parent company guarantees are often required in today's construction market. The cost of a bond will depend on the suggested wording, duration, value and creditworthiness of the contractor. The quotation for a bond is usually expressed as a percentage of the contract value per annum, and extends into the defects liability period. This means that a bond for a twelve-month contract with a twelve-month defects liability period will require twice the annual cost.

Overheads and profit must be added to the total given in line 11; dayworks and provisional sums (undefined) are added afterwards because they carry their own overheads.

Overheads and profit

There are three main stages in adjudicating a tender. It is management's responsibility to:

1. Understand the nature and obligations of the work.
2. Review the costs given in the estimate, and if necessary adjust the costs for market conditions and errors.
3. Add to the estimate sums for general overheads and profit.

Overheads and profit should be evaluated separately because they are calculated in different ways for different purposes.

Item 12. The term 'overheads' relates to off-site costs which need to be recovered to maintain the head office and local office facilities. Items to be covered include:

- Salaries and costs to employ directors and staff;
- Rental fees, rates and maintenance of offices, stores and yard;
- Insurances;
- Fuel and power charges;
- Cars and other vehicles costs for office staff;
- Printing, stationery, postage and telephone;
- Advertising and entertainment;
- Canteen and consumables;
- Office equipment including computers;
- Finance costs and professional fees.

These charges are compared with turnover to arrive at an overhead percentage. Most organizations will know the figures for previous years, but both overheads and turnover should be predicted for the future when the project is under way.

Unfortunately, when work is scarce and turnover drops, contractors look for ways to reduce tender mark-ups. The temptation is to reduce the amount for overheads at a time when they are rising in proportion to turnover. The alternative is to win less work and suffer large losses. Another solution would be for some contracts to make a greater contribution to head office costs than others.

Item 13. The profit figure is a combination of discounts and additional profit required by management. Long gone are the days when discounts could be thought of as a small reserve fund. In a competitive market all discounts are taken out before a small profit margin is added, to help win the work.

The profit calculation is the responsibility of senior managers (and ultimately the directors). In fact it is not strictly a calculation but a view or hunch about what margin would give the maximum profit for the company with the likelihood of winning the contract.

There are, nevertheless, some important issues which must be considered before a tender can be completed, these include:

1. The desire to win the contract; perhaps to increase turnover or the job might be the first in a number of similar schemes.
2. Whether the project will involve contractor's finance; the cashflow calculations will show the net finance needed.
3. The effect of winning the contract on the present workload; is there sufficient turnover to meet the company's objectives and are the company's resources being used efficiently?
4. Knowledge of the client and his consultants; the attitude and competence of other parties can have an impact on the smooth running of a project.
5. The local market conditions; consider the strength of competition for the type of construction in the area (this is often the single most important criterion for choosing a winning profit margin).
6. An evaluation of previous bidding performance; knowledge is gained of profit margins by an examination of results from previous tenders.
7. There is a theory that contractors may be influenced by the client's budget; this target is found from 'intelligence' information. It may be given in the preliminary invitation to tender, or deliberately released by the client to keep the price down; in practice this seldom changes the profit margin.

An examination of the risks to be borne by the contractor may be considered at this stage. These can be divided into technical and contractual:

1. Technical risks are dealt with by defining construction methods before

costing the work. If the cost of failure is high in relation to the value of the project, it may be possible to insure against the loss, or increase the control. When uncertainties have been assessed they are priced by adding lump sums which are a proportion of the possible losses. Contractors usually take an optimistic view on the unknown events which can plague a construction project, hoping that costs can be held within the site and off-site overheads.

2. Contractual risks are those imposed by the form of contract and additional obligations forming part of the agreement. The most common problems arise from failure to finish by the date for completion. If management feel that the contract period could be exceeded, they should consider adding a sum equal to the liquidated damages which might be claimed.

Where the tender appears to exceed the sum which would win the contract, there are two refinements which can be used:

1. Re-examine the suppliers' and sub-contractors' quotations for any evidence that lower prices may be available after the main contract is awarded, and
2. Consider different profit margins for direct work and those for which sub-contractors will be responsible.

Finally, once the overheads and profit are settled, the amounts can be put on the summary form. All that remains is to add daywork (item 14) and undefined provisional sums (item 15) to arrive at the overall total, and, where applicable, add professional fees to produce the tender figure.

When the adjudication meeting is over it is important for the estimator to return to his desk and check all the figures on the summary sheet once again. Even better, he must enter the estimator's adjustments and adjudication changes into the bills of quantities and ensure that the bill total is the same as the agreed tender figure. This fully adjusted bill will form the basis of a set of allowances for the construction team and can be adapted for presentation to the client.

Tender submission and results

SUCCESSFUL TENDERS

UNSUCCESSFUL TENDERS

An assessment of tender performance

Introduction

How do you submit a tender? The answer was simple: filling in the tender form and making sure that it was delivered on time. Today, however, the winning of a contract can be more than simply giving the lowest price. The style of presentation, for example, is important for design and build contracts, and

225

method statements are commonly required by management contractors. There is, of course, a balance to be achieved between what the client wants and appearing to be too clever; different clients have different expectations.

The contractor needs to consider the criteria the client will use for selection. These can be:

1. Price; will the lowest price alone be the basis for selection?
2. Time; will a programme show the client that the contractor has thought about how the job can be finished on time, or ahead of time?
3. Allocation of money; will the way in which money is distributed in the priced bills help or irritate the client?
4. Method statement; would the client wish to know the methods to be adopted before accepting the offer?
5. Safety and quality; does the client expect a statement of safety or quality showing how the contractor will manage this particular contract?
6. Construction team; is the contractor proposing to supervise the job with experienced staff who will work as an effective team with the consultants?
7. Presentation; how important is an accurate, well-presented offer?

The follow-up to a tender can affect the outcome – the estimator will contact the client soon after submitting a tender, not only to find the result but also to ensure that there are no questions arising out of the submission.

Completion of priced bills

The Tender Summary form produces a tender sum which must be transferred to the bills of quantities for submission. If a priced bill of quantities has to be submitted with the tender, then the way in which money is spread in the bill should be decided at the adjudication meeting. The contractor often changes the actual breakdown of prices in a priced bill of quantities, to:

1. Produce a reasonable cashflow from interim valuations.
2. Apportion monies in a way which the client will find acceptable.
3. Increase the money set against undermeasured items and decrease the price of overmeasured items.

The example adjudication forms given earlier show that the first bill total was £664 705 before the preliminaries were added and any adjustments were made. The tender sum arrived at after adjudication was £698 437. If the bill was inked in using the first pricing level, the amount remaining for preliminaries would be:
    £698 437 − £664 705 = £33 732

This is not the true preliminaries sum but is the amount needed to bring the bill total up to the tender sum. If the parties agreed to proceed on this breakdown (and they probably will) a part of the preliminaries, overheads and profit would remain in the measured work portion of the bill. In particular, the items to be carried out by domestic sub-contractors would carry a considerable mark-up.

Taking the example a stage further, if the estimator had the use of a computer and had made the tender adjustments before inking in the bill for the client, the breakdown would be in line with the tender summary below:

Measured work and provisional sums:		529 413
(including £620 fluctuations on labour)		
Preliminaries		
Labour	12 960	
Plant	37 890	
Materials	5 780	
Sub-contracts	3 300	
Staff	37 619	
Fluctuations	1 675	
Water	650	
Insurances	5 000	
Bond	1 050	
O/heads & Profit	18 000	123 924
Provisional sums and Daywork		45 100
Tender total		£698 437

The client should not be surprised to see this large sum for preliminaries because it is based on the true allowance. If the contractor anticipates a problem with this breakdown, he can move some money either:

1. Into 'safe' items in the measured work portion of the bill, looking for work which will be carried out early in the contract (safe items are those which appear to be measured correctly or are judged to be undermeasured at tender stage), or
2. By using a computer system to add a percentage to all the rates in a bill of quantities.

The preliminaries total should be broken down in the bill of quantities with sums for fixed and time-related items. A surprising number of contractors ignore this breakdown and prefer to insert a lump sum in the collection; they assume that if their tender is the lowest, more details can be submitted to meet the needs of the quantity surveyor (or engineer) for valuation purposes. The contractor also knows that if there are some small queries raised by the client then his tender is (probably) being considered for acceptance.

When contractors are tendering in a competitive market in which work is scarce, they know that their bids must be close to the predicted cost of carrying out the work, with little mark-up. Sometimes tenders can be slightly below cost. The contractor, in taking a calculated risk on how the contract will turn out, may price some items in a way which appears to be inconsistent. As an example, assume that a bill of quantities has two equal amounts in items for breaking out rock: one in reducing the site levels and the other in excavations for drains. Contractor A priced both items at £18.00/m³ and contractor B priced the rock in open excavations as nil and in drains at £36.00. The overall effect on the tender sum was the same but contractor B had discovered a serious under-measurement in the drainage bill; he was therefore hoping the drainage bill would be remeasured and valued at the higher rate. This might appear to make sense but, as many contractors have learned to their cost, plans can go wrong. If the quantity of rock in open excavation increased substantially, the contractor would suffer a serious financial loss.

Tender presentation

Most tenders are submitted on a pre-defined form which has the effect of standardizing the offers and discouraging exclusions, alternative bids and other qualifications. For traditional procurement methods, the tender presentation normally includes the form of tender and a covering letter.

With design and build projects, the contractor submits his tender in the form of 'the Contractor's Proposals'. This is the contractor's response to the Employer's Requirements and is explained in detail in Supplement Number Two to the CIOB Code of Estimating Practice. The common elements of a design and build offer are: drawings to illustrate the proposals, a detailed specification and a tender sum broken down into its major elements. In order to avoid confusion later, the employer should stipulate the form and extent of information needed. The Contract Sum Analysis should be adequate for both valuing work executed and changes after the contract is awarded.

Instructions to tenderers should include a date, time and location for submitting a tender. The contractor is responsible for presenting the documents by the time given and in some cases may be permitted to send a tender by facsimile transmission, followed up by first-class post. Contractors rarely submit their tenders early for two reasons:

1. They might receive a lower quotation from a sub-contractor or supplier which could improve the bid, and

2. A tender price cannot be communicated to a competitor in time for him to better the price.

Clearly, if a form of tender is required, the contractor must enter the price in the space provided and ensure that the document is signed and dated by a person authorized to act for the tenderer. The estimator must carefully check the instructions to tenderers for any other documents to be submitted with a tender. Many public organizations would not expect to see additional documents, but sometimes ask for an outline programme and method statement. Contractors commonly attach a letter to the form of tender and priced bills of quantities, but are careful not to add statements or conditions to their offer which might be seen as qualifying the tender.

There are times, of course, when qualifications are unavoidable. The following examples show how a contractor may have no choice but to bring matters to the client's attention either before submission or in the tender:

1. A contractor may decide that the wording of a performance bond is unacceptable.
2. Having examined all the resources needed for a job, the contractor may find that the contract duration is too short.
3. Late amendments can impose extra responsibilities which the contractor is unable to resolve.
4. The contractor may wish to add to the list of three named sub-contractors following the failure of one of the firms on the list.

There is a theory (which is not sensible in a competitive market) that all such problems can be resolved by 'throwing money at them'. A short duration, for example, can be overcome by adding liquidated damages for the expected overrun. The Code of Procedure for Single Stage Selective Tendering suggests that the tenderer should tell the client if there are any matters needing clarification as soon as possible and preferably not less than 10 days before the tenders are due. If the tender documents need to change, the tender date may be extended. The Code takes a strong line on qualified tenders by stating that qualifications should be withdrawn otherwise the tender may be rejected. With this in mind, qualifications are sometimes written in general terms so that the contractor can delay his decision on which issues he wants to qualify. Typical statements used are:

1. 'During the tender period, we identified some savings which can be brought about by small technical changes . . .'
2. 'We would need to clarify some of the contractual matters before entering into an agreement, but do not expect this to affect our price . . .'

229

Another approach, more common in civil engineering contracts, is to submit an alternative tender. In this way a contractor is able to comply with the tender conditions by submitting a 'clean bid' and at the same time reveal an alternative offer which usually reduces the construction costs with only minor specification/contractual changes. An alternative tender may also be the vehicle to propose a shorter duration, submit a programme and impress the client with technological expertise.

There is a growing practice of submitting company brochures, technical literature and other publicity material with the tender. This 'window dressing' is often unnecessary. The client is more interested in the price and approach to *his* job; in any case, the company profile has been examined at the pre-selection stage.

The letter accompanying the tender can be used to confirm the amendments to the tender documents received during the tender period. The basis of the offer is, after all, the tender documents and all amendments received by and not sent to the contractor before the tender date. The main rule for this letter is to keep it short, no more than one page. The rules of letter writing should be applied to all correspondence, particularly during the tender stage, when for some clients this is the first business contact. In writing to his clients, the estimator must remember that every letter sent to the client, or his advisers, is selling the company.

Vetting of tenders

If bills of quantities are not required with the tender form, the contractor who has submitted the lowest bid is asked to submit his priced bill of quantities for examination (and adjustment where errors are found). The unsuccessful contractors should be told immediately that their tenders were unsuccessful. Once the contract has been let, all the tenderers should be notified of the results so that they can measure their performance against others in the industry. The results must remain confidential before a contract is made because the lowest-bidding contractor could negotiate higher prices if he knows the tenders made by his competitors. There have also been instances of higher tenders being reduced below the lowest price received on the tender date.

An examination of a contractor's bill of quantities will reveal different kinds of errors. Some errors should be corrected using the Code of Procedure for Single Stage Selective Tendering, Alternatives 1 or 2. Alternative 1 gives the contractor the choice of standing by his tender or withdrawing it. Alternative 2 gives the contractor the opportunity to confirm his offer or amending it to correct genuine errors. The term 'genuine errors' is not defined in the Code, but normally means:

1. Mathematical errors caused by mistakes in multiplying quantities by rates or totalling pages;

2. Patent errors in pricing such as pricing hardwood joinery at the same rates given for softwood earlier in the bill, or pricing steel reinforcement at a rate per kilogram where the unit is tonnes;
3. Inking-in errors are usually simple to amend because the summaries will be correct even if an individual rate has been entered wrongly.

Some patent errors can be difficult to spot because the contractor may have his own commercial reasons for distributing the money in a certain way. A common difficulty is the pricing of similar items at different prices. A quantity surveyor may wonder why concrete in a ground floor slab is priced at £55.35 per m³ in the workshops and at £65.65 in the office area of a factory development. There are many logical reasons for this apparent mistake. It could be to do with continuity of work, perhaps the latter case involved a high extra cost for part load charges from the concrete supplier. If the workshops were programmed for the end of the contract, the quantity surveyor might think the contractor had 'front-loaded' the bill to produce an early income. The contractor cannot be required to change the rates but the quantity surveyor may not be able to recommend the tender to the client. If the contractor has priced the bill in such a way that considerable sums of money are overpaid at the beginning of a project, the client might be at risk if the contractor fails to meet his obligations.

Post-tender negotiations and award

Each tenderer will want to know the result as soon as possible in order to plan the construction phase of successful bids, or file away the documents and redeploy resources involved in unsuccessful ones. The direct approach is usually the most productive. A telephone call to the person carrying out the vetting process is often enough to know whether the tender documents can be filed and the computer files backed up onto floppy disks.

The next stage for the lowest tenderer could be meeting the client, or his advisers. This is often necessary before an award can be made. The matters discussed are mainly financial and contractual although methods can be important. Any errors or discrepancies in the bill of quantities can be resolved at the meeting and contractual details can be discussed and agreed. This allows both parties to understand their obligations before the formal agreement comes into effect.

The contractor should be represented by the estimator and senior construction staff. The estimator should ask for an agenda and a list of those attending. The agenda will allow him to brief his team in advance and take relevant documents to the meeting. The list of client's representatives and advisers is important. The contractor will try to respond to questions with staff who have the necessary specialization. Above all, an estimator must avoid the situation where he alone enters a room where all the consultants and client's representatives are assembled confidently expecting to get the best deal for the client.

There are some pre-award meetings where the estimator may not be the best person to lead the contractor's team. The approach must be robust with a firm commitment to carrying out the work to a high standard and on time. Unfortunately, estimators often get bogged down in detail and have been known to highlight small errors in the documents or be pessimistic about aspects of the programme. If this happens, a senior manager can present a wider view and suggest positive remedies which have been successful on other projects. Above all, the client must have confidence and believe that the contractor can carry out the work with a willingness to solve problems and work closely with the client's team at all times.

Tendering performance and analysis of results

There are several ways in which the performance of an estimating department can be measured. The simplest method would be to count the number of successful bids compared with the number of tenders submitted. Figure 18.1 shows the cumulative ratio of tenders to contracts won in eight-months. This is a crude technique which does not help the firm to improve its tendering performance.

If a contractor's business strategy is to increase turnover, then a simple graph showing the value of contracts awarded would be useful. Figure 18.2 shows the value of contracts won, with the total for tenders submitted. This graph is made more effective by showing the performance related to a target set by the board of management.

The CIOB Code of Estimating Practice offers suitable forms for recording the results and subsequent analysis of a tender. Figure 18.3 illustrates three methods which can be used to produce a ratio analysis for a tender; each contractor will pick the method which helps him to evaluate his tender performance. The underlying principle is that, in general, cost category ratios of similar jobs remain approximately the same. If there are significant changes in these ratios, the estimator should be able to explain the reasons for the deviations at the adjudication meeting.

Column A of Fig. 18.3 has been calculated with each element being expressed as a percentage of the total tender sum. This is the least refined measure because the elements are being compared with a figure which includes overheads and profit. Column B shows the elements expressed as percentages of direct costs excluding PC and provisional sums. This has the advantage of removing the sums fixed by

CB CONSTRUCTION LIMITED

TENDER PERFORMANCE   1994

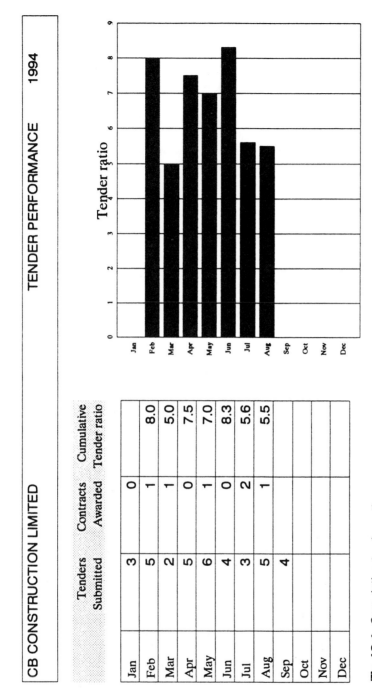

	Tenders Submitted	Contracts Awarded	Cumulative Tender ratio
Jan	3	0	
Feb	5	1	8.0
Mar	2	1	5.0
Apr	5	0	7.5
May	6	1	7.0
Jun	4	0	8.3
Jul	3	2	5.6
Aug	5	1	5.5
Sep	4		
Oct			
Nov			
Dec			

Fig 18.1 *Cumulative tender ratio*

CB CONSTRUCTION LIMITED

TENDER PERFORMANCE   1994

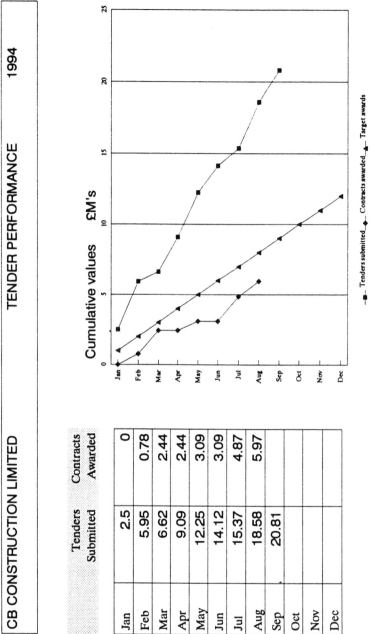

	Tenders Submitted	Contracts Awarded
Jan	2.5	0
Feb	5.95	0.78
Mar	6.62	2.44
Apr	9.09	2.44
May	12.25	3.09
Jun	14.12	3.09
Jul	15.37	4.87
Aug	18.58	5.97
Sep	20.81	
Oct		
Nov		
Dec		

Fig 18.2 *Cumulative value of tenders and awards*

CB CONSTRUCTION LIMITED		Tender analysis	Project name :	Fast Transport Limited
			Tender date :	Jul 1994
			Tender no :	T354

Tender summary			Column A	Column B	Column C
Description		£	% of 13	% of 6	% of 9
1 Own measured work Labour		115627	16.6	22.8	18.3
2 Plant		26030	3.7	5.1	4.1
3 Materials nett		143091	20.5	28.2	22.7
4 Domestic sub−contractors nett		220935	31.6	43.5	35.1
5 Price fluctuations		2295	0.3	0.5	0.4
6 Direct costs		507978		100	
7 Project overheads		104249	14.9		16.5
8 Overheads, profit and discounts		18000	2.6		2.9
9 Tender less fixed sums		630227			100
10 PC Sums − Nom sub−contractors		19500	2.8		
11 PC Sums − Nom suppliers		3610	0.5		
12 Provisional sums and contingencies		45100	6.5		
13 TENDER		698437	100		

Fig 18.3 *Alternative methods for tender ratio analysis (relates to data given in Tender Summary form)*

the client but fails to bring the project (site) overheads into the calculation. Perhaps the best solution would be to include project overheads and take out all sums set by the client (see column C).

The next stage is to examine bid results in more detail. Figure 18.4 lists the tenders submitted and shows the effect of subtracting the sums set by the client. The percentage over lowest bid gives a measure of the margin by which the job was lost, and the percentage over mean bid provides the contractor with a guide to the deviation from the average prices set by other contractors. As a rough rule of thumb, many contractors would feel confident in their estimating performance if

235

CB CONSTRUCTION LIMITED		Project name:	Fast Transport Ltd
		Tender date :	Jul 1994
		Tender no :	T354

Tender list	Tender	Tender less set sums	% over lowest bid	% over mean bid
1 Baldwin Bros	764780	696570	14.4	6.2
2 Javelin Construction	707990	639780	5.0	−2.4
3 Barry and Hardcastle	677250	609040 L	0.0	−7.1
4 CB Construction	698437	630227	3.5	−3.9
5 Wessex Contracting	718415	650205	6.8	−0.8
6 Newfield Building	756410	688200	13.0	5.0
7 Simpson	743382	675172	10.9	3.0
Mean tenders	723809	655599	7.6	0.0

Sums set by client	
PC Sums – nom subs	19500
PC Sums – nom suppliers	3610
Prov sums	45100
Total	68210

Margin lost by	£21,187.00

Fig 18.4 *Tender results*

their percentage over mean bid fell within the range ± 10%. Figure 18.5 gives a summary of tender results for a three-month period.

It would be unwise to change a bidding strategy using a small amount of data. The analysis of performance must start with:

1. A period of consistent net pricing.
2. A steady tendering policy in terms of adjustments to direct costs and the percentage mark-up.
3. A determination to obtain tender results from consultants and clients.

CB CONSTRUCTION LIMITED		Summary of tender results				1994

	Project title	Tender date	No. of Tenderers	Rank	% over lowest bid	% over mean bid
T351	Renton Cakes	Jun 94	3	1	0.0	−2.3
T352	Warland Access Road	Jun 94	6	5	11.0	4.9
T353	Retail units, Swindon	Jun 94	6	1	0.0	−3.1
T354	Fast Transport Limited	Jul 94	7	2	3.5	−3.9
T355	Star Tyres	Jul 94	5	2	5.0	−3.8
T356	Cinema Roof	Jul 94	6	4	12.3	2.2
T357	Fire Station	Aug 94	6	3	6.7	1.2
T358	Railway workshops	Aug 94 .	7	1	0.0	−7.7
T359	British Coal underpinning	Aug 94	13	5	15.0	1.0

Fig 18.5 *Summary of tender results*

Bidding Strategy

A bidding strategy can be defined as a broad framework of methods and timing to achieve stated objectives. It is interesting to note that in military terms, the word 'strategy' means the skilful management of an army in such a way as to deceive an enemy and win a campaign. In business the stated objectives can sometimes be achieved by deceiving the opposition but principally the specific objective is to be successful in winning contracts at prices which would allow the organization to carry out work profitably. A tendering strategy can be developed as a statement of aims, as follows:

1. To identify a suitable market in terms of type of work, size of contracts and geographical location.
2. To develop a reputation for quality and speed of construction within economic limits.
3. To secure stated targets for turnover.
4. To evaluate the company's performance, and compare with that of its competitors.
5. To compare the financial performance of a project with the costs predicted at tender stage.

A contractor can improve his tendering efficiency with better marketing, only accepting invitations to tender which meet the guidelines for type of work, size and location of the project.

Finally, it is worth considering a different approach; that is, to ignore the tender levels set by your competitors. Produce pricing levels which are right for your company and avoid playing the market. If prices fall below what you consider to be an economic level, look for other markets where margins can be preserved.

Action with the successful tender

19

Introduction

Gone are the days when the award of a contract signalled the end of the estimator's contribution to a project. There are two important tasks to be performed: first, helping the construction team with procurement and technical advice during the mobilization period, and second, producing cost information which will form the budget for the job. The transfer of information has been improved considerably with the introduction of computers. There are many estimating packages available which can be used to produce tender allowances and later assist during the construction phase to control sub-contractors' payments and produce valuations for the client. The main advantage of using a computer is that an estimator can adjust his estimate to take account of the decisions made at the adjudication meeting and make post-tender adjustments which are sometimes agreed with the client following the vetting stage. For close financial control of a project, the site manager should be trusted with the detailed budget; he can then control the resources efficiently and contribute to a simple feedback system.

Information transfer

The estimator must take great care to produce accurate information in a form which is simple to understand. A checklist can be used to ensure that the handover information is complete, such as:

1. Correspondence with the client.
2. Form of tender.
3. Priced bill of quantities.
4. Tender drawings and specifications.
5. Contractor's bill of allowances (adjusted rates).
6. Tender stage programme.

7. Pricing notes and resource schedules.
8. Project overheads schedule (adjusted following adjudication).
9. Sub-contractors' and suppliers' quotations.
10. Any other findings or assumptions which might include temporary works drawings, photographs, site layouts, minutes of meetings with specialists and the client, technical literature and site visit reports.

The item which needs the most effort is the contractor's bill of allowances because this gives the site manager the fully adjusted rates for the work. The changes made during the tender period must be applied to all the rates affected. It is not satisfactory, for example, to say to a site manager that the budget for concrete is as given in the client's bill less 5% which was an additional discount taken during the adjudication meeting plus a 20% addition to the waste allowance for concrete poured against the sides of excavations. No site manager has the time to trace such changes made by the estimator.

There has been a cynical view in the industry that an estimator's figures should not be made available to construction staff on site. There are three main reasons for this attitude:

1. The construction team should get the resources for the job at the lowest possible rates and not just beat the tender target figures.
2. There may be a security problem if the rates are open to view on site; a competitor may steal an advantage and sub-contractors could be upset if they had sight of the real allowances.
3. There may be problems with interpreting data, and staff can become confused when confronted with bills produced for different purposes.

A more subtle reason might be that some organizations like to adjust the financial targets as more is known about the job. As an example, a contracts manager might decide that a concrete pump will not be necessary for placing concrete and hopes the site manager has not seen this provision in the tender. This clearly shows a lack of confidence in the site manager and could adversely affect the financial control of the project.

The important point is that the site manager must appreciate the cost implications of the decisions made on site, particularly with the use of direct labour, the costs of materials and equipment and the value of sub-contracted work. The information provided by the estimator can be used:

1. To set bonus levels.
2. To produce forward-costing data.
3. To quantify resources for planning exercises.

4. To examine the financial performance of a contract through historical costing methods.
5. To compare the final building costs with the tender budget.

The person on site will perform better if trusted with important information; he is then more likely to accept greater responsibility for the financial success of a project and contribute more effectively to the evaluation of future projects.

The extract from a bill of allowances given in Fig. 19.1 was produced by a computer package which was used by the estimator during the tender stage and the quantity surveyor during construction. It gives the true allowances to the site manager (columns a to d) and the rates submitted to the client in column e. The rates given to a client will differ from the tender allowances for many reasons. These include:

1. Changes are made to rates after the submission bill is inked in. For example, a site might be found for tipping surplus material which reduces the rate for disposal. In extreme cases, the bill can be inked-in before the adjudication meeting. This would lead to many changes.
2. Some of the margin may be included in the rates submitted to the client.
3. Money may have been moved to ensure early payments or to take advantage of mistakes in the bill, the true rates must be given to the site manager for cost control purposes.

There are, in effect, two bills of quantities – the client's bill for valuation purposes and the contractor's bill for costs control.

Before work can start on site the construction manager will bring together all those associated with the contract. This internal pre-contract meeting is an opportunity for the estimator to introduce the scheme to the construction team, personally explain the contents of the handover package, and expand on the methods of construction, resources and organization used as the basis of the tender. The estimator will be given certain duties at this meeting, mainly to do with the transfer of information described earlier and checking the contract documents, and he may be asked to take part in negotiations with sub-contractors appointed at the start of the project.

The CIOB Code of Estimating Practice recommends procedures for checking contract documents. The estimator must ensure that the contract documents received for signature are identical to those which were used as the basis of the tender, and any amendments issued by the client during the tender stage have been correctly incorporated. Correspondence confirming post-tender negotiations may also be included if it helps to clarify the basis of the agreement.

CB CONSTRUCTION LIMITED		Bill of Allowances	Project name:			Fast Transport Ltd		
			Contract no:		94022	Date:		Aug 1994

Bill Ref:	Short description	Quantity	Site Allowances						Margin/	Client's bill	
			Lab (a)	Plt (b)	Mat (c)	Sub (d)	Rate	Tot	adjustmt	Rate (e)	Tot
3/9 A	Filling to excavations	18 m3	1.70	1.65	8.85		12.20	219.60	1.00	13.20	237.60
B	Filling to soft spots	100 m3	1.20	1.40	8.85		11.45	1145.00	0.55	12.00	1200.00
C	Level & compact exc	224 m2	0.45	0.25			0.70	156.80		0.70	156.80
D	Dust blinding and compact	2890 m2	0.12	0.12	0.32		0.56	1618.40		0.56	1618.40
E	Grout under baseplate	21 m2	24.00		2.00		26.00	546.00		26.00	546.00
F	Plain conc 15N trench	76 m3	12.20		36.60		48.80	3708.80	5.20	54.00	4104.00
G	Plain conc 15N isol founds	4 m3	13.00		36.60		49.60	198.40	5.40	55.00	220.00
	Page total	3/9	2081	573	4939			7593	490		8083
3/10 A	Reinf conc 35N founds	52 m3	11.45		39.60		51.05	2654.60		51.05	2654.60
B	Reinf conc 35N isol founds	5 m3	12.65		39.60		52.25	261.25		52.25	261.25
C	Reinf conc 35N col casing	12 m3	26.30		42.25		68.55	822.60		68.55	822.60
D	Fabric D49 wrapping	40 m2	1.40		0.75		2.15	86.00	0.70	2.85	114.00
E	Fabric A393 in slab	2765 m2	0.85		3.30		4.15	11474.75	0.40	4.55	12580.75
	Page total	3/10	3381		11919			15299	1134		16433

Fig 19.1 *Extract from contractor's bill of allowances*

Feedback

Feedback is the weakest part of the estimating function. In practice, an estimator receives information about the actual costs of construction in a haphazard way and usually hears, through a third party, about underestimated costs – rarely will he be told about high budgets. So why are companies slow to set up procedures which would ensure that feedback information is available to estimators? The following problems are often quoted:

1. Feedback information is historical.
2. Each project is different from the next.
3. Financial performance is determined by the effectiveness of the site management team.
4. An estimator uses constants which are not always job-specific.
5. A feedback system would be expensive to implement.
6. Market prices can change quickly with little notice.
7. Confidential information is not available to site staff.

There might be a middle course by which a company can report on certain aspects of a job in progress, as follows:

1. The actual costs associated with project overheads could be written in a spare column on the schedule produced by the estimator.
2. Individual investigations can be carried out to find the actual waste of high-value materials.
3. The average cost of employing certain categories of labour could be compared with the all-in rate used at tender stage.
4. The value of sub-contracts (and major material orders) which have been let can be entered on a comparison sheet (see Fig. 19.2). This would give management, and estimators, evidence of the buying margins which are available in the current market.
5. For small repetitive jobs, where detailed feedback is needed, an extra column could be inserted in the bill of allowances for the eventual costs to be added to each item. This is an obvious application for a computer using a tailor-made package or a spreadsheet program.

   The relative importance of these investigations will depend on the estimator's need for information and the size of each contract. The benefits are that future estimates become more reliable and more accurately reflect the cost of construction work.

CB CONSTRUCTION LIMITED Sub—contracts placed			Project name : Date : Contract no:		Fast Transport Limited Mar 1995 94022	

Trade	Tender allowance		Contract placed			
	Name	net value	Name	net sum	difference	
1 Roof covering	Grange Roofing	16400	Grange roofing	14760	1640	
2 Windows	Aliframe	25500	Westpoint Windows	23455	2045	
3 Plasterwork	Macdonald	23450	Macdonald	25670	−2220	
4 Partitions	Port Drylining	20050	Port Drylining	20050		
5 Joinery	Robin Joinery	34000	Robin Joinery	29550	4450	
6 Ceilings	Wignall Hampton	19800	Shrimpton Ceilings	18600	1200	
7 Painting	T & G Jackman	12300				
8 Floor coverings	ABA Furnishings	12300				
9 Surfacing	Gatwick Plant	4200				
10 Landscaping	no quote	7600				
11 Mechanical	Moss and Lamont	39100	Hutley Engineers	37145	1955	
12 Electrical	Tate Electrics	31900	Tate Electrics	31900		
	total	£246,600	total		£9,070	

Fig 19.2 *Comparison of sub-contracts placed with sub-contract allowances*

244

20 Computer-aided estimating

The paperless office

Aims of computer-aided estimating

At the beginning of the 1980s contractors and computer experts were excited by the introduction of desktop computers which could react interactively with the operator using a visual display unit. Unfortunately, there were two wrong assumptions made about estimating with computers: first, that the computer could replace the estimator's skills in building up unit rates and second, that large databases of items could be the foundation of an estimating system. By the early 1990s the situation has changed; the estimator now builds up his rates by putting together resources in any way he wishes, and systems are sold with databases of prices as optional extras. The keyword is 'flexibility'.

The computing debate has raised questions about the role of the estimator and whether estimators should change their methods to conform to computing techniques or whether computers should be used to mimic the way estimators have worked in the past. What appears to have happened is that estimators have developed their computing skills, not just in using estimating systems but adopting spreadsheet and database packages where appropriate; and the software specialists are beginning to respond to the needs of estimators with more flexible systems. There is still a market for the large database of standard items, probably in the bill-production phase whether created by the private QS or contractor.

So why was computing so successful in accounting, payroll and word-processing, and not in pricing construction work? It could be that these applications had clear aims for their development and required fewer skilled operators to input the information. The estimator needs to make decisions throughout the pricing stage mainly because each project is different. The standard labour and plant outputs which he has in mind are adjusted to suit the circumstances. These circumstances might include many variables such as distance from compound, ground conditions, depth below ground, plant available on site, item sizes, quality of workmanship, quantity (scope of work), degree of repetition, access and so on. The estimator clearly needs skills which are judgement based as well as the ability to work consistently at mechanical processes. The software must allow the estimator to exercise his judgement particularly when he is building up his rates for each job.

The general aim of computer-aided estimating is to provide the estimator with a kit of tools which will enable him to save time and exercise his personal judgement within a given framework, with reasonable scope for flexibility and user-ingenuity. The software needs to benefit first, the estimator in his role as the person who calculates the total net cost of the project, and second, those who have to make decisions based on the estimator's reports and allowances.

The wider benefits of computer-aided estimating include the opportunities for contractors to gain a commercial advantage over their competitors. Spreadsheets

246

and databases are commonly used for early cost estimates and cost planning, where a model for a proposed scheme can be drawn up by reference to the elemental costs of previous jobs. The contractor is well placed to offer this advice. Tender presentations are much more sophisticated today. Clients expect contractors to demonstrate their capacity to work not only to a fixed price but also to a resourced programme and quality plan. Increasingly, clients call for these details at tender stage. Fortunately, this information can be produced quickly with word-processing, desktop publishing and graphics packages. But the real advantage is the facility to edit text and graphics in order to produce polished presentations.

The best buying advice given to anyone entering the computer marketplace is to choose the software first and then find the hardware to run it. Finding the right software is notoriously difficult because the benefits are intangible, especially in relation to your particular needs. So the answer must be to consider what you want to do; in other words, start with what you know and work towards what you need to know. The following list gives some of the potential uses for a computer:

Cost planning	(SS, DB or PK)
Tender register	(DB)
List sub-contractors and suppliers	(DB or WP)
Enquiry letters	(DB or WP)
Resource price lists	(DB, SS or PK)
Calculate all-in rates	(SS or PK)
Produce standard bills for repetitious work	(SS or PK)
Bar schedules	(SS or PK)
Rate build-ups	(SS or PK)
Extend and total bills of quantity	(SS or PK)
Lists of company staff costs and plant	(SS or PK)
Calculate costs of fluctuations	(SS or PK)
Adjust for late quotations	(SS or PK)
Calculate/plot cash flow analysis	(SS)
Reports for management	(SS or PK)
Adjust individual resources	(PK)
Adjust/distribute mark-up on rates	(SS or PK)
Gross bill for client	(SS or PK)
Bill of allowances for construction	(SS or PK)

Key:
DB   Database program, or package with database facilities
SS   Spreadsheet (limited scope in comparison with a specialist package for estimating, but more flexible)
WP   Word-processor package
PK   Specialist package for computer-aided estimating

Computer-aided estimating packages

Estimators and their managers have not been slow to recognize the potential of computers to increase the efficiency of the estimating process, but during the 1980s were disappointed with the systems on offer and the problems of implementation. It could be said that contractors and software providers were looking at the problems from opposite perspectives. Contractors wanted software which would mimic their methods when in fact the systems were being developed to make best use of the hardware and programming techniques available. The result was a false start in computing because computers have not matched users' expectations.

Some argue that software providers have tackled the estimating challenge in the wrong way by creating huge databases of work items in order to mirror all the possible items that could be envisaged in a bill of quantities. This 'price library' approach conflicts with the way estimators work. A good estimator can calculate the cost of an item of work far quicker manually than using a list of rates. Printed bills of quantities often have thousands of work items which need to be matched with the coded descriptions in a computer library. This tedious task must be done by an experienced estimator who is also able to recognize and deal with rogue items. The library method can delay the pricing of printed bills of quantities. The counter-argument is that where a contractor carries out work in a certain sector of the construction market many of the work items repeat. The library can be used to build up a series of standard bills of quantities which would need minor changes on each estimate. Perhaps the best use of the library is the production of bills where the tender is based on a specification and drawings. The database of work items prompts the estimator with descriptions and guide prices. The items can be taken off in any order because the software will put the items into the sequence recommended by the standard method of measurement, and print trade bills for sub-contract enquiries.

By the end of the 1980s attempts were being made to answer many of the criticisms, as follows:

1. The software reverted to a 'shell' arrangement whereby estimators could create a database for the items in the current job only. They no longer attempt to build a comprehensive database for future projects (some standard bills will be kept if similar projects are expected).
2. The complex price build-ups have been replaced by simple resource calculations which have a common method of entry for all items of work, for example:

Amount	×	Resource type	×	Cost	=	Rate
0.14 (hours)		carpenter		8.00	=	1.12
1.00 (m)		100 × 50 wallplate		1.10	=	1.10

The individual resources can be priced during the entry of the work item or all the resources can be dealt with separately.

3. A system of menus is used to guide the estimator around the system and context-sensitive help screens bring relevant advice to the user whenever he needs it. 'Context-sensitive' refers to the way in which software will display helpful information on the command that is about to be carried out or may advise a user who is uncertain about how to continue.
4. More powerful computers are available with the latest microchips and high-capacity storage devices with fast access times.
5. Menus, help screens and utilities can be called up on the screen using windows of information. Switching between windows allows the estimator to undertake various subsidiary tasks while he is working on the estimate.

There are many packages advertised in the construction press, but few independent test reports. The claims made for each package are similar and very difficult to confirm. A short demonstration by a salesperson is not a reliable way to evaluate a system because deficiencies will be glossed over. It is important to ask about the facilities being offered by various suppliers, and write down which facilities will be of most benefit (see Fig. 20.1). It is unlikely that the software will meet all the needs of a company but some are more flexible in use than others. For example, there are some systems which store their data in DBase format which can be output to other databases or sent to a word-processor to produce high-quality presentations.

Estimating systems may not be able to offer all the features listed in Fig. 20.1 and many estimators will not need all the features. An optical character recognition system will not be needed, for example, by an estimator who inputs bill items by reference to page and item numbers; a direct labour organization may not need facilities for adding labour-only sub-contractors; and a small company might not need a multi-user system.

Clearly, estimating systems store a great deal of information about a project which can be linked to project planning, buying, valuation and accounting packages; in many ways this is what makes the use of computers worth while. The information produced at tender stage may need to be changed once a tender is successful. It would be unsafe to order materials, for example, from tender stage bills of quantities because the drawings may have changed. A project quantity surveyor would wish to insert item descriptions where the estimator did not.

Main characteristics	Subsidiary characteristics
Hardware requirements	Compatibility
	Memory requirements
	Multi-user systems
	Optical character recognition
	Digitizer facilities
Price	Initial
	Up-grades
Method	Database of standard text and build-ups
	User-definable database
Help	Free on-line help
	Context-sensitive help screens
	User manual
Reports	Outputs fixed by software
	Output designed by estimator
Specific facilities	Sub-contract comparisons
	Labour-only rates substitution for direct labour rates
	Files compatible with Dbase or ASCII
	Database runs in 'RAM disk'
	On-screen calculator
	Sort items by trade or user codes
	Windows
	Detection of unpriced items
	Nested work assemblies
	Mouse driver
	Recalculation/reporting speeds
	Checking procedures

Fig 20.1 *Features checklist for estimating packages*

There is a danger that people can be too concerned with generating data for its own sake. It is commonly said that 'information is an organization's most valuable asset', but do we always need so much information? Estimators are resigned to the fact that most of their reports are ignored once a job is won. There is, nevertheless, a clear advantage to be gained by making pricing data available to construction staff. The computer will produce the properly priced (net) bill, the commercially priced (gross) bill, and any number of (package or trade) bills for negotiations with sub-contractors.

It is worth remembering that computers reduce the clerical effort and reproduce data in a sorted form, but, above all, cannot do anything that you cannot do manually (in time).

General-purpose software

A desktop computer can handle a vast range of programs and carry out thousands of different tasks. Most of the software can be purchased off-the-shelf from either a specialist producer or a retailer. There are two main types of software: applications programs and systems programs. The kinds of programs available in each category are:

Applications programs
1. General-purpose packages (see Fig. 20.2) (e.g. word-processors, spread-sheets and databases).
2. Specialist packages (e.g. estimating, accounts and expert systems).

Systems programs
1. Operating systems (e.g. MS-DOS and UNIX).
2. Utility programs (e.g. to reinstate a file erased in error).

One of the most difficult decisions for computer users is whether to buy specialist software which has been tailored for a particular application or use general-purpose programs such as spreadsheets and databases. For example: if an estimator wants to produce a small bill of quantities he can use a word-processor, database or spreadsheet. He could alternatively decide to use a specialist bill-production package. The main functions of general-purpose software are as shown in Fig. 20.2.

Word-processors, spreadsheets and databases are the most popular products in the general-purpose software market, closely followed by computer-aided design and project planning.

Word-processors

Word-processing is an electronic means of making the task of writing easier than using a typewriter or pen and paper. Text is entered via the keyboard, just like a typewriter, and appears on the monitor before being printed. It is then possible to check, modify or store the text before it is sent to the printer for a paper copy. The ease with which changes can be made to a document has helped estimators to prepare professional-looking presentations without being competent typists. Any word-processor software can reduce the time required to produce a complex typed document, because initial drafts can be edited and corrected so easily. The production of standard documents for enquiry letters, method statements and design/build presentations is much quicker since the introduction of computers and correspondence can be personalized with little effort.

TYPE	TEXT	CALCULATIONS	FILING/ SORTING	GRAPHICS
Wordprocessors	★★★	★		★
Spreadsheets	★	★★★	★	★★
Databases	★	★	★★★	★
Graphics	★			★★★
Desktop publishing	★★★			★★★
+Integrated software	★★★	★★★	★★★	★

+integrated software is a combination of several
different applications with the facility for the user
to transfer data from one package to another

KEY :

★   = limited functions

★★★ = extensive functions

Fig 20.2 *The main funtions of general-purpose software*

Main features of a word-processor package
Add and delete text
Combine separate documents
Check spelling
Move and copy blocks of text
Justify left and right margins
Automatic page numbering
WYSIWYG – 'what you see (on the screen) is what you get' (on your printed page)

Other useful features
Mathematical functions
Transferable file formats so that a file can be read by another word-processor or desktop publishing system
Thesaurus
Mail-merge (merge database of names and addresses)
Line drawing important for flow charts and bills of quantities
Mouse support

Full WYSIWYG with alternative text fonts
Full-page preview allows the user to see a picture of the whole page before printing
Mathematical and technical symbols
Full range of printer drivers including control of the fonts offered by different printers
Macros allow the user to define a list of keypresses
Clipboard to store a block of text in memory for future use
Word count

It is not easy to decide which program to buy. Most well-known packages are suitable for day-to-day use and offer many of the features listed above. Often the safe answer is to choose the market leader because the files will be compatible with many other systems and staff prefer to develop skills which will be widely accepted.

The estimator can make use of the basic word-processor features for letters, method statements, quality statements and safety plans, and has particular needs for specification writing and bill production. In building, the National Building Specification is available as simple text files from NBS Services Ltd. The subscriber will find that most computer systems are supported by this service, in terms of both disk medium and software. Small bills of quantities are easier to produce using spreadsheet software but where a large unpriced bill is to be written, a word-processor will accept many more pages of information. Normally, bill production packages use a database method which overcomes the size limitations.

Spreadsheets

The key ingredient which has led to the widespread acceptance of the personal computer, as something more than a clever typewriter, is the spreadsheet. It is simple to use, and does not try to change the way people undertake their calculations. Most of the repetitious work of an estimator could be computerized without the help of a programmer. The immediate benefit is the fast recalculation of look-up charts of the most important rates. Estimators can test the effect of changing parameters, often referred to as 'what if' calculations.

A spreadsheet can be visualized as a large piece of paper divided into boxes, called cells. The user can refer to any cell by giving its column letter and its row number. Figure 20.3 shows a standard spreadsheet layout with the entry cursor at cell B3. With the cursor placed over the required cell, the estimator can enter NUMBERS, TEXT or FORMULAE. To add a column of figures, the cursor is placed in the answer cell and a formula is entered; in Fig. 20.3 the formula in cell

	A	column B	C	D	E
1				21	
2				43	
row 3			Total =	64	
4					
5					
6					

Fig 20.3 *Spreadsheet layout*

D3 would be D1 + D2 or if the whole column needed a total in cell D6 the formula would be sum(D1.D5). The formula is not shown on the spreadsheet, but the answer is.

The power of the spreadsheet is seen in the way cells can be linked by formulae; the overall effect of a change anywhere on the sheet can be seen immediately. What you see on the screen is a 'window' displaying only part of the whole sheet; if the entry cursor is moved to the edge of the screen then a new portion of the spreadsheet comes into view. Although some manufacturers claim a maximum sheet size of 256 columns wide and 9900 rows deep, it is unlikely that a desktop computer would have enough working memory to handle such huge amounts of data. A large spreadsheet could be produced by expanding the size of the computer's random access memory, using multiple spreadsheets or adopting efficient memory-management routines such as disabling the 'undo' feature.

It is becoming easier to enter text into a spreadsheet. Most of the leading packages have elementary editing facilities although they cannot be compared with the sophistication of modern word-processors. Other non-spreadsheet features have been added, such as database functions and some useful graphics for printing graphs or just high-quality presentations of lists of results. If a very wide spreadsheet has been created, then it can be difficult to get a copy reproduced on a printer. A clever utility program is available which rotates the spreadsheet by 90 degrees as it prints. Modern spreadsheet packages include printer drivers which can compress a large spreadsheet so that it can be printed by a laser printer by scaling down the size of the text font. Spreadsheet design can be speeded up by using macros. The macro is a time-saving method to automate a series of instructions. For example, a commonly used set of key strokes can be assigned to a single key.

Since the early 1980s many spreadsheet programs have been developed. The pioneering software was VisiCalc, which set the pattern for others to follow. Lotus 1-2-3 has become the market standard because it was the first to integrate the early spreadsheet designs with database features and powerful graphics. Supercalc has also been popular with its more straightforward approach to data entry and advanced features for data management. The following guidelines can be used to select a suitable package:

1. Spreadsheet capacity – if the estimator needs to create spreadsheets which occupy many cells then the software should either support an extended memory board or have the ability to place information on disk when available random access memory is filled. Although modern personal computers have a large working memory, spreadsheet programs account for much of it, sometimes more than half of the 640K which is now standard on MS-DOS systems.
2. Spreadsheet speed – will the software support a maths co-processor which will speed up the recalculation of large spreadsheets?
3. Range of features – mathematical, statistical and financial functions; macro commands for pre-programmed operations; undo to correct the last entry to the spreadsheet; powerful formatting and copying commands, sideways printing, and security against unauthorized access.
4. Integration – of data management, graphics and ease of editing text and numerical entries.
5. File linking – allows the user to draw information from other spreadsheets stored on disk. This is used by estimators for creating a summary page which brings together the totals from individual bill sections.
6. 3D capabilities – multiple spreadsheet pages can be loaded into memory as a single file and all linked formulae on all pages will be updated when a recalculation is performed.
7. Ease of use – may be subjective, but today the software should support the use of a pointing device such as a mouse, and help should be available from any level of operation (context-sensitive help windows).

There have been several attempts to provide alternative approaches to spreadsheet programs but most improvements have been made in the presentation capabilities. These include WYSIWYG and other desktop publishing features such as mixing graphics, lines and text; page preview; and full support of laser printer facilities. A formula goal-seeking command is available for when you know the answer you want a formula to give, but not the values needed to reach it. A typical application for this would be in trying to adjust a cost plan to arrive at a client's budget. Contractors also find this facility useful in adjusting the rates in bill of quantities by a percentage which would give the required total.

Someone new to spreadsheets should start with those applications which lend themselves to the tabular presentation of data. The most elementary would be:

1. Look-up charts for reinforcement, brickwork, drainage and fixing ironmongery.
2. Small bills of quantities for composite items such as manholes, kerbs and simple house extensions.
3. Domestic sub-contractors' quotation analyses.

After a little practice, the following could be attempted:

1. All-in hourly rate calculation.
2. Plant rate build-ups.
3. Bills of quantities for standard house types.
4. Look-up charts for more complex rates, such as formwork and disposal of surplus excavated material.
5. Elementary cost planning.

More advanced applications include:

1. Cashflow forecasts.
2. Project overheads schedules.
3. Bills of quantities for uncomplicated commercial and industrial buildings, and plant foundations.
4. Reinforcement schedules.

The examples shown in Figs 20.4 and 20.5 were produced by a groundworks sub-contractor to create quick look-up charts for pricing bills of quantities. The item highlighted in Fig. 20.4 shows the total rate for formwork to beams where the fix and strike time is 1.70 hours and the estimator expects four uses of the shutter. The estimator can change any of the data at the top of the page and the total rates change within seconds. These applications show that a spreadsheet can closely mimic the traditional methods used to produce rates, but do so at much greater speed with clear presentation but, as with all computer methods, still need careful interpretation by an experienced estimator.

Anyone who is used to dealing with figures will soon be charmed with the power of such sophisticated software, and will be able to test various theories to arrive at the best condition or price. There are, however, many dangers awaiting the unwary estimator. The problems arise when:

256

FORMWORK	Project :	LIFEBOAT STATION	DATE:	JUN 94

ENTER BASIC DATA

Carpenter (all-in rate/hr)	8.00	
Labourer (all-in rate/hr)	6.50	
Plywood cost per sheet	17.00	
Softwood cost per m³	185.00	

Percentage additions

LAB for travelling	0.00	
, for fluctuations	5.00	
for o/heads & profit	10.00	
for discount to MC	2.50	
Total mark up on labour =	1.18	
MAT/PLT for fluctuations	5.00	
for o/heads & profit	10.00	
for discount to MC	2.50	
Total mark up for MAT/PLT =	1.18	

ASSUMPTIONS

Waste %			9.00 ..%
Consumables	making		0.08 ..xcraft rate
	fixing		0.07 ..xcraft rate

	Founds	Soffit	Beams	Walls	Columns	
make	0.80	0.50	1.00	0.95	0.80	..hrs/m2
timber	0.04	0.04	0.05	0.05	0.04	..m3/m2
plant	0.44	4.00	1.10	1.05	0.85	..£/m2

Lab assist handling	(hrs)	0.15 ..hrs/m2

HOURS TO FIX AND STRIKE		1.20	1.30	1.40	1.50	1.60	1.70	2.00
ONE USE	Founds	33.93	35.55	37.18	38.80	40.43	42.28	45.79
	Soffits	35.78	37.46	39.14	40.81	42.49	44.42	47.99
	Beams	38.34	40.10	41.86	43.61	45.37	47.40	51.05
	Walls	37.89	39.63	41.37	43.12	44.86	46.87	50.51
	Columns	35.44	37.11	38.78	40.45	42.12	44.03	47.59
TWO USES	Founds	25.26	26.49	27.71	28.94	30.16	31.39	34.64
	Soffits	28.12	29.41	30.69	31.97	33.25	34.53	37.88
	Beams	28.12	29.40	30.68	31.96	33.25	34.53	37.87
	Walls	27.83	29.11	30.39	31.66	32.94	34.22	37.55
	Columns	26.33	27.58	28.83	30.07	31.32	32.57	35.86
THREE USES	Founds	21.66	22.71	23.76	24.81	25.86	26.91	29.86
	Soffits	24.94	26.03	27.11	28.19	29.28	30.36	33.34
	Beams	23.87	24.94	26.02	27.09	28.16	29.23	32.20
	Walls	23.66	24.73	25.80	26.87	27.94	29.01	31.97
	Columns	22.55	23.61	24.67	25.73	26.79	27.85	30.80
FOUR USES	Founds	19.77	20.72	21.67	22.61	23.56	24.51	27.35
	Soffits	23.28	24.22	25.17	26.12	27.06	28.01	30.85
	Beams	21.64	22.59	23.54	24.48	25.43	26.38	29.22
	Walls	21.46	22.41	23.36	24.31	25.25	26.20	29.04
	Columns	20.56	21.51	22.46	23.41	24.35	25.30	28.14
FIVE USES	Founds	18.56	19.50	20.45	21.40	22.35	23.29	26.13
	Soffits	22.20	23.15	24.10	25.05	25.99	26.94	29.78
	Beams	20.21	21.16	22.10	23.05	24.00	24.95	27.79
	Walls	20.06	21.00	21.95	22.90	23.85	24.79	27.63
	Columns	19.29	20.24	21.18	22.13	23.08	24.03	26.87
SIX USES	Founds	17.75	18.70	19.64	20.59	21.54	22.48	25.33
	Soffits	21.49	22.44	23.38	24.33	25.28	26.23	29.07
	Beams	19.26	20.20	21.15	22.10	23.04	23.99	26.83
	Walls	19.12	20.07	21.01	21.96	22.91	23.85	26.70
	Columns	18.44	19.39	20.33	21.28	22.23	23.17	26.02

Fig 20.4 *Example of a spreadsheet template for formwork*

DRAINAGE	Project :	LIFEBOAT STATION	Date :	JUN 94

ENTER BASIC DATA (rates/hr)		Assumptions (can be changed)			
			0−1.5m	1.75−2.75	3m+
Margin %	10				
Labourer	6.50				
NWRA	0.15	Rate of dig m³/hr	12	8	6
Excavator	4.90	Earth support cost	1.80	2.10	2.50
Fuel	3.25	Grade and ram cost	0.40	0.45	0.50
Operator plus rate	0.35				
Pump	1.20	Distance to tip	....	8	km
Compactor	1.20	Tip charges	....	1.50	/m3
16t lorry (all−in)	16.00	(or work to tip)			

			Pipe sizes (mm)				
DRAIN EXCAVATION			100− 150	225− 300	375− 450	525− 600	675− 750
DEPTH	width	>	0.60	0.70	1.10	1.25	1.45
0.50			3.93	4.26	5.56	6.05	6.70
0.75			5.77	6.23	8.10	8.80	9.73
1.00			7.60	8.21	10.64	11.55	12.76
1.25			9.44	10.18	13.17	14.30	15.79
1.50			11.27	12.16	15.71	17.04	18.82
	width	>	0.70	0.80	1.20	1.40	1.60
1.75			16.60	17.82	22.68	25.12	27.55
2.00			18.92	20.31	25.84	28.61	31.37
2.25			21.25	22.80	29.00	32.10	35.20
2.50			23.57	25.28	32.15	35.59	39.02
2.75			25.89	27.77	35.31	39.08	42.84
	width	>	0.80	0.90	1.30	1.50	1.70
3.00			35.44	37.81	47.28	52.02	56.75
3.25			38.36	40.92	51.16	56.29	61.41
3.50			41.28	44.03	55.04	60.55	66.06
3.75			44.19	47.14	58.93	64.82	70.71
4.00			47.11	50.25	62.81	69.08	75.36

Fig 20.5 *Example of a spreadsheet template for drainage excavation*

1. The estimator who builds a spreadsheet model fails to produce a foolproof design, or carries out inadequate checks.
2. Another estimator inadvertently changes data in a model or erases a formula by entering a number in a formula cell.

Often a user is unable to spot mistakes in his own spreadsheet; he is more inclined to believe the results when they are presented on a computer printout. The effect of using inaccurate answers from such calculations could be ruinous in a tender. The following guidelines will help to prevent such errors:

1. Start by planning the general requirements with a sketch showing the labels and layout of the spreadsheet. The optimum size of a spreadsheet will depend on available memory and the data an operator will want to see either on the screen or close to the edges.
2. Adopt a modular approach where the layout will include separate identifiable sections, such as:
    (a) An instructions portion.
    (b) An area where the user can change data freely.
    (c) The results or summary section.
    These sections could be created in different (but linked) files, on pages of a 3D spreadsheet, or in different parts of the screen display.
3. Protect formulae from accidental erasure or amendment by putting them in the results area of the spreadsheet. It is also possible to make a formula secure by using the password-protection feature found on most versions of the program.
4. Check that the numbers representing money are not only to two decimal places for display purposes but also for subsequent calculations in the model. The reason for this check is that although the spreadsheet has been instructed to show two decimal places it usually keeps a more accurate number in the computer's memory.
5. Carry out simple checks using data from previous manual systems. Other estimators could be asked to test the model to find any bugs or misleading instructions.

Above all, where more than one person is going to use the spreadsheet, keep the design simple.

Spreadsheet programs now offer a safe choice for organizations which recognize the need to introduce computers to their staff cautiously and at low cost. They are powerful in the rapid production of pricing notes which are usually in a tabular form. The estimator can create spreadsheets which are an amalgam of his expert knowledge by holding the production information which he has collected.

Once a format has been created (and saved) for a particular purpose, it is referred to as a 'template'. This is because the layout and formulae will usually be fixed but the data variables can be changed.

Most estimators will be inspired to extend the range of applications and build a valuable selection of templates to suit their own needs and methods of working. On average, an estimator will take about six hours to build a template if he is familiar with the software.

Databases

A database is a computerized filing system, the electronic equivalent of a card index or filing cabinet. A database program allows the user to file information and retrieve it in many ways. Perhaps the most common example is a list of names and addresses. When a list has been built up it is called a *file* of data. A file can be displayed as a list on the screen, sorted into order, searched for individual pieces of information and printed. Figure 20.6 shows a typical *record* which is the basic building block of a database. Each record contains a number of *fields* of information, such as the name field or postcode field.

A simple database can be set up in few hours. The user first designs the layout for the record screen defining the fields which are needed. Second, the program can be used to produce printouts in different formats. For example, an address file could be printed with a list of names in the first column and telephone numbers in the next. Before printing lists of data, the program could be asked to sort all the records into alphabetical order; or search for all the suppliers in a particular town or district. The program can also create and store a standard letter so that names and addresses can be merged with text to prepare letters for a selective mail shot, for example. More powerful packages offer much greater scope for manipulating and analysing data and include their own programming language used to develop more sophisticated applications.

The range of applications is large in an estimator's office, from elementary filing tasks to the complex manipulation of data. Examples include:

1. Address lists for suppliers and sub-contractors.
2. Drawing registers.
3. Tender registers.
4. Marketing information.
5. Cost-planning data.
6. Bill production and pricing.

There are many national databases available to construction organizations willing to attach their computers to the telephone system, using a modem. The user is expected to pay an annual subscription and an additional fee each time the

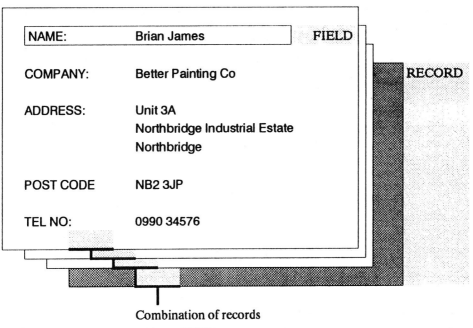

| NAME: | Brian James | FIELD |

COMPANY: Better Painting Co — RECORD

ADDRESS: Unit 3A
Northbridge Industrial Estate
Northbridge

POST CODE NB2 3JP

TEL NO: 0990 34576

Combination of records
makes a FILE

Fig 20.6 *Terms used in a simple database programme*

database is accessed. There is a growing market for these databases to be available on a read-only compact disc (CD-ROM) which is a much more convenient medium for a desktop computer system.

Planning

Many of the programmes made by estimators can be produced with paper and pencil and a calculator. An estimator's programme is rarely needed during the construction phase but will be used until the mobilization stage to assess the project overheads and undertake a cashflow analysis. Where a planning engineer produces the preliminary programme, the continuity is likely to be stronger and the programme may well form the basis of a control/monitoring document.

There are several reasons for using project planning software to aid manual systems at tender stage:

1. It may be easier to keep complex projects under control.
2. The program will calculate resource loadings and plot resource histograms.

3. The critical activities can be identified, which might affect the workforce levels used for pricing.
4. High-quality charts and clear presentations could impress a client at tender stage.

There are many planning packages available today; some such as 'Hornet' give a wide range of powerful features and flexibility to implement the software in different ways. There are others which offer mouse-driven input to powerful software at lower cost. 'PowerProject', for example, has excellent presentation features and yet is easy to use. The estimator will look for software which will show the overall project duration, highlight critical activities and allow labour and plant resources to be analysed.

Graphics

Desktop micros use graphics for three main purposes:

1. Business graphics – the graphical representation of data drawn from information in a spreadsheet or database.
2. Desktop publishing – the in-house design and publication of leaflets and forms, usually with the aid of a laser printer.
3. Computer-aided design (CAD) usually output to a high-quality plotter.

It is estimated that 60% of the total cost of design work is attributable to freehand drawing and detailed line drawing. For this reason CAD has become popular with consultants and contractors alike. The production gain depends on the type of work being designed and the degree of repetition; the estimate for construction is in the range 2:1 to 15:1. The main problem is that the initial investment can be very high. On top of the cost of workstations and plotters must be added a large investment in training. Training and lost productivity can cost as much as hardware and software.

There are many software packages available for small micros, to do simple line drawing, rendering and 3D modelling. Industry leader Autodesk produce the AutoCAD design package with an architectural add-in which incorporates the symbol and layering conventions of BS 1192. Most packages offer features such as the ability to scale drawings up and down in size, adding text and dimensions, keeping shape libraries, and of interest to an estimator is the ability to attach prices to certain items within a drawing.

Shareware

Choosing an estimating system for an organization is a difficult task which fills some purchasers with fear. A few system providers lend their prospective clients a system for on-site evaluation and a small number sell a demonstration disk. There is another group of software, called shareware, mainly written in the USA, which is distributed on a 'try before you buy' basis. Shareware software may be copied and distributed freely, but if after trying the program you find a need for it, you are asked to contribute towards the cost of development by registering your copy with the author. Because shareware authors have low advertising budgets and few employees the registration fee is low and always lower than comparable commercial software. For educational establishments, the attractions of shareware are obvious. Students can copy the software which they use on the college computers for use at home, and for a small fee can register their program and receive a manual and the latest update. Great care is needed, though, where software is exchanged freely because this is the prime cause of viruses infecting computers.

For general-purpose use there are some excellent shareware databases, spreadsheets and word-processors available which can be evaluated for the cost of a floppy disk. The quality of specialist estimating (bidding) software is questionable for construction in the UK. Many of the programs have been written for specialist trades such as sheet metal workers or electrical sub-contractors. When you find a comprehensive program, such as PC-Estimator, you will not be able to use the cost database because it uses the American CSI format and prices. This will be of interest to anyone looking into the cost of building in the United States, and all the rates can be changed if you wish to tailor the system to your needs.

Accurate Estimator is not a comprehensive package and the comparison made in Fig. 20.7 shows that it cannot be considered as competing with Manifest. Perhaps shareware is best used as an introduction to computing to get a feel for computer methods and to avoid the risk of losing money. The most serious omissions from Accurate Estimator are:

1. Sub-contractors' quotations cannot be entered against a group of items and the software will not insert rates in the tender.
2. There are no material schedules. Rates for materials are entered into standard build-ups or when the estimate is produced.
3. The facility to add a waste factor is missing: presumably waste must be part of the basic material rates.

These are significant problems. There are, however, some useful functions which are not always available in full-price packages:

System/supplier	*Manifest* 16a Molivers Lane Bromham Bedford MK43 8JT 02302 4801	*Accurate Estimator* *(also known as PC Estimator)* CPR Inc, 3195 Adeline Street Berkeley California 415 654-8338
Hardware support	Digitizer OCR	Measure-Mouse
Compatibility	MS-DOS	MS-DOS
Multi-user systems	Yes	No
Method	User-definable database	User-definable database
Additional facilities	Valuations BOQ production	Built-in calculator Bid results analysis
Memory requirements (min)	512K RAM	448K RAM
Ease of use	Good	Simple
Links with other software	Various	ASCII File dBaseIII compatible
Help screens	No	No
Look-up windows	Yes	Yes
User manual	Yes	Yes
Sub-contract comparisons	Yes and automatic entry into estimate	No
Built-in reports	Yes	Yes
Demonstration program	Yes £50	Yes PC Estimator as shareware £2.50
Training courses	If requested	No
Telephone support	Yes	Yes (to registered users) in USA
Training in-house	Yes	No
Price	£1750	£100

Fig 20.7 *Comparison of a typical full-price and shareware package*

1. Context sensitive look-up windows (see Fig. 20.8) allow the estimator to inspect the contents of the database. This means that the user does not need to remember codes when building up a tender.
2. The software will run on a lap-top computer without a hard-disk drive (may be slow and will limit the size of file).
3. The database files are compatible with dBaseIII and can be transferred to Lotus and other spreadsheets. The reports can be output as text files for use with any word-processor.
4. An on-screen calculator is available from all screens (Fig. 20.9) and the estimate total can be viewed at any time (Fig. 20.10).

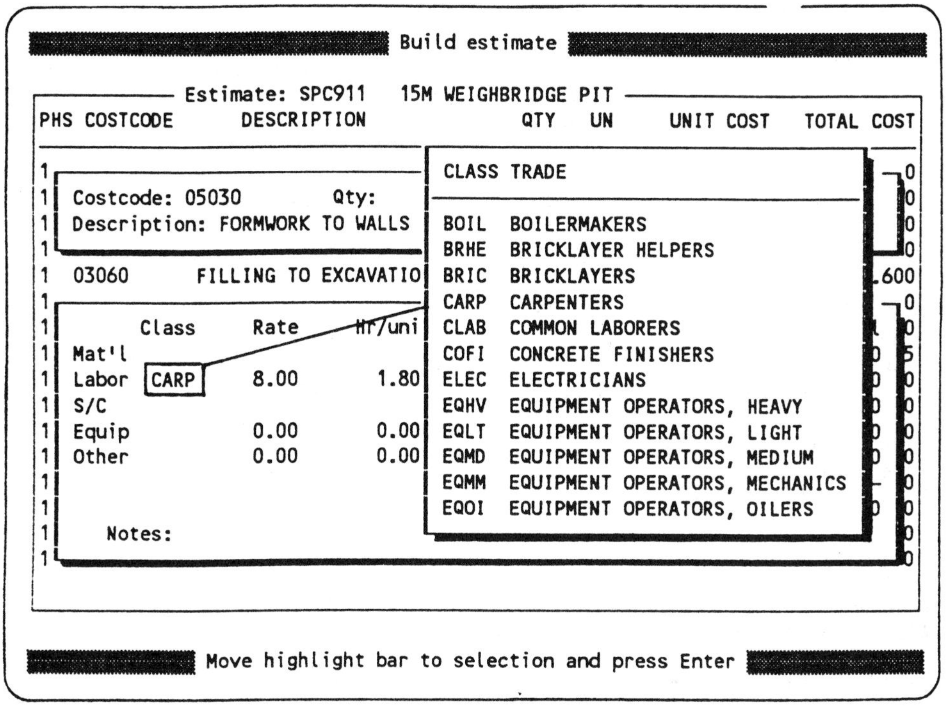

```
╔══════════════════════════════════════════════════════════════╗
▓▓▓▓▓▓▓▓▓▓▓▓▓▓▓▓▓▓▓▓▓▓  Build estimate  ▓▓▓▓▓▓▓▓▓▓▓▓▓▓▓▓▓▓▓▓▓▓

─────────── Estimate: SPC911   15M WEIGHBRIDGE PIT ───────────
PHS COSTCODE      DESCRIPTION          QTY   UN   UNIT COST  TOTAL COST

1 ─────────────────────────────      CLASS TRADE                      0
1  Costcode: 05030     Qty:          ─────────────────────            0
1  Description: FORMWORK TO WALLS      BOIL  BOILERMAKERS              0
1 ─────────────────────────────      BRHE  BRICKLAYER HELPERS         0
1  03060      FILLING TO EXCAVATIO    BRIC  BRICKLAYERS             .600
1                                     CARP  CARPENTERS                 0
1        Class     Rate     hr/uni    CLAB  COMMON LABORERS
1  Mat'l                              COFI  CONCRETE FINISHERS        5
1  Labor  [CARP]   8.00     1.80      ELEC  ELECTRICIANS
1  S/C                                EQHV  EQUIPMENT OPERATORS, HEAVY
1  Equip           0.00     0.00      EQLT  EQUIPMENT OPERATORS, LIGHT
1  Other           0.00     0.00      EQMD  EQUIPMENT OPERATORS, MEDIUM
1                                     EQMM  EQUIPMENT OPERATORS, MECHANICS
1                                     EQOI  EQUIPMENT OPERATORS, OILERS
1     Notes:
1

▓▓▓▓▓▓▓▓  Move highlight bar to selection and press Enter  ▓▓▓▓▓▓▓▓
╚══════════════════════════════════════════════════════════════╝
```

Fig 20.8 *Look-up window giving choice of labour. The look-up feature is context-sensitive and can be used to enter data directly from the list*

Spreadsheets distributed as shareware are popular, particularly those such as As-Easy-As which started out as Lotus lookalikes. They don't all have the sophisticated features such as 3D or typeset quality output, but this does not mean that they should be dismissed. What you want from a spreadsheet depends on what you want to solve. The shareware packages all carry out the financial and statistical tasks well but often lack the control over screen presentation and printing which is well supported by the full-price systems.

Hardware

A basic computer system is made up of the computer itself, a keyboard, a monitor and storage devices including at least one floppy disk drive and a hard-disk drive. An estimator will need to add some 'peripheral' devices to suit the demands of the

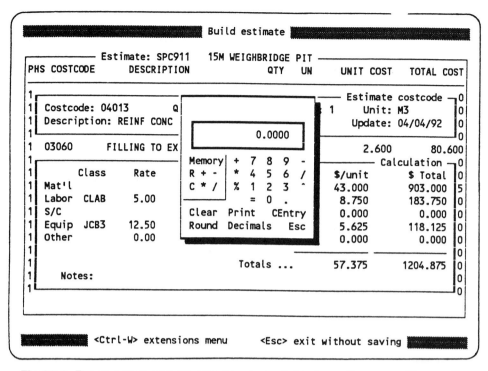

Fig 20.9 *The on-screen look-up calculator is available from all screens. The result of a calculation can be posted to a numeric field*

software he is going to use and the need to produce quality presentations. A 'desktop' micro is characterized by the 'three-box' design; the system unit, monitor and keyboard. Lap-top computers now offer the power of a desktop micro including 3.5-inch disk drive, hard disk and high-resolution LCD screen.

A typical specification for a desktop computer is shown in Fig. 20.11. Most of these facilities are available in a lap-top computer, but the price would be significantly more if you wanted a colour LCD display and such a large hard-disk capacity.

The most important part of a computer system is the software, because it is this which dictates what can and cannot be done. On the other hand, now that there is a vast range of software which can run on a standard microcomputer, people prefer to select the micro first. Of the distinct groups of computer available today, the most popular is the IBM compatible. Following the success of the IBM PC much good-quality software was produced and high-quality compatible equipment has been manufactured by companies which have built up a reputation to match that

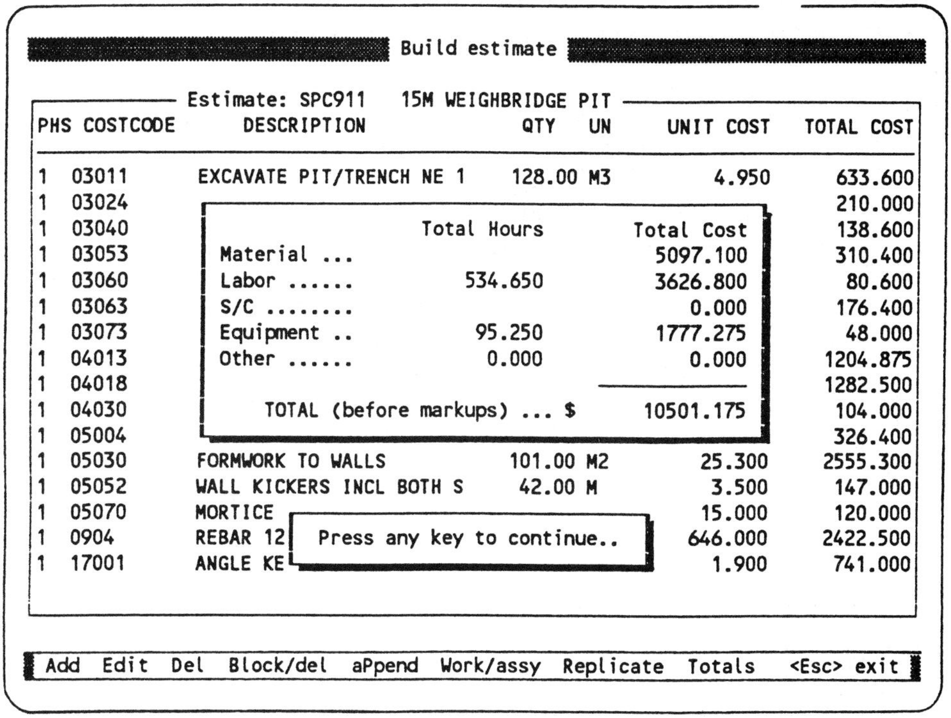

```
████████████████ Build estimate ████████████████████

┌──────────── Estimate: SPC911   15M WEIGHBRIDGE PIT ─────────────
│ PHS COSTCODE       DESCRIPTION       QTY   UN    UNIT COST   TOTAL COST
│
│1   03011   EXCAVATE PIT/TRENCH NE 1   128.00 M3      4.950     633.600
│1   03024                                                       210.000
│1   03040   ┌──────────── Total Hours        Total Cost        138.600
│1   03053   │ Material ...                     5097.100        310.400
│1   03060   │ Labor ......     534.650         3626.800         80.600
│1   03063   │ S/C ........                       0.000        176.400
│1   03073   │ Equipment ..      95.250         1777.275         48.000
│1   04013   │ Other ......       0.000            0.000       1204.875
│1   04018   │                               ──────────        1282.500
│1   04030   │  TOTAL (before markups) ... $   10501.175        104.000
│1   05004   └────────────────────────────────────────         326.400
│1   05030   FORMWORK TO WALLS         101.00 M2     25.300     2555.300
│1   05052   WALL KICKERS INCL BOTH S   42.00 M       3.500      147.000
│1   05070   MORTICE ┌───────────────────────────┐  15.000      120.000
│1   0904    REBAR 12│ Press any key to continue..│ 646.000     2422.500
│1   17001   ANGLE KE└───────────────────────────┘   1.900      741.000
│
└──────────────────────────────────────────────────────────────

████ Add  Edit  Del  Block/del  aPpend  Work/assy  Replicate  Totals  <Esc> exit ████
```

Fig 20.10 *'Build estimate' screen showing totals window*

of IBM. Figure 20.12 shows the minimum arrangement for general-purpose applications. As computers continue to become more powerful, and new software is written to make use of the increased power, a computer buyer should consider getting an up-gradable PC. Many manufacturers now offer PCs designed to allow the board carrying the processor to be easily removed and replaced with faster chips.

The optional devices, called 'peripheral' because they operate outside the central processing unit, will depend on what the computer will be used for, what software will be run on it, and to some extent the PC configuration such as the type of display. The most important output device is the printer which will be either dot-matrix or laser. As the price of laser printers is falling, they are superior to dot matrix on every count: print quality, speed (up to 12 pages per minute) and versatility. Laser printers take data from the PC and build up an image of the page in their own internal memory. When the image is complete, the page is printed by firing a laser onto a photosensitive drum. The process is then similar to that of a photocopier.

1	Intel 80486 microprocessor with inbuilt maths co-processor
2	128K RAM cache memory
3	4 MB RAM expandable to 32MB on board
4	1.2MB 5.25" floppy disk drive
5	1.44MB 3.5" floppy disk drive
6	210MB hard disk drive with 15ms access time
7	CD ROM drive
8	15" super VGA colour monitor with resolution up to 1024 x 768 in 256 colours
9	Clock with battery back up
10	200W power supply
11	5 expansion slots free
12	1 parallel and 2 serial ports
13	Microsoft compatible mouse

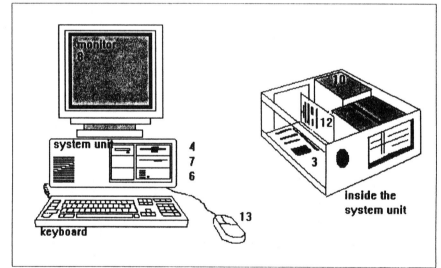

Fig 20.11 *Typical specification for a desktop computer*

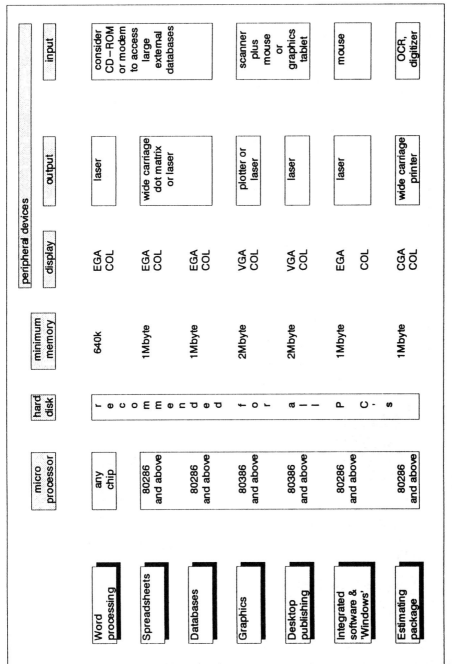

peripheral devices

software	micro processor	hard disk	minimum memory	display	output	input
Word processing	any chip		640k	EGA COL	laser	consider CD–ROM or modem to access large external databases
Spreadsheets	80286 and above		1Mbyte	EGA COL	wide carriage dot matrix or laser	
Databases	80386 and above	recommended for all PCs	1Mbyte	EGA COL		
Graphics	80386 and above		2Mbyte	VGA COL	plotter or laser	scanner plus mouse or graphics tablet
Desktop publishing	80386 and above		2Mbyte	VGA COL	laser	
Integrated software & 'Windows'	80286 and above		1Mbyte	EGA COL	laser	mouse
Estimating package	80286 and above		1Mbyte	CGA COL	wide carriage printer	OCR, digitizer

Fig 20.12 *Selection of hardware*

269

There has been much interest in input devices which can save estimators time. The obvious solution would be to send contractors bills of quantities in a standard format, on a floppy disk, or as a file sent electronically along a telephone line. In the meantime efforts have been made to use digitizers and optical character readers.

A digitizer is used to take measurements from a drawing. Some of the features of a digitizer are:

1. Automatic scale adjustment – useful if the software with a digitizer can accurately compensate for reduced or photocopied plans based on known dimensions found on a drawing.
2. Calculation modes to measure irregular shapes and angles.
3. Compatibility with existing equipment and spreadsheets.
4. Automatic transfer of measurements to quantity field in database.
5. Switchable between imperial and metric scales.

The digitizer software is usually menu-driven using an overlay window over the taking-off program. An audit trail can be kept of every measurement so the take-off can be checked manually.

If an estimator wants to input a printed bill of quantities into an estimating system he will usually enter the item references and a shortened description. Sometimes a full copy of a bill is input using an optical character reader. An OCR is a scanner which can recognize characters on a piece of paper and reproduce the text in an estimating program, database or word-processor. There is some concern in the construction industry that time must be set aside for checking the computer version of the bill and the purchase of an OCR device is not worth while until the benefits are clear and the technology improves.

An existing technology, which has been in widespread use in the USA for several years, is the CD-ROM which is an acronym for Compact Disc Read-Only Memory. The Compact Disc technology is the same laser technique used in audio music discs. A 12 cm disc can hold up to 600 Mbytes of data – the equivalent of over 800 average capacity floppy disks. The term Read-Only Memory means that the user cannot modify the CD to add or change data. The potential of CD-ROM technology is exciting, not only because of the vast amounts of data which can be stored on a single disc, but the data can represent anything from basic text files and graphics to moving video images and sound, making CD-ROM a powerful information medium. To read a CD-ROM disc the computer must be fitted with a CD-ROM drive, which can be fitted externally alongside a PC or internally into the front of a PC system unit. There are even portable drives made for portable PCs.

Implementation

Computer-aided estimating is not a single program or technique but the development of opportunities provided by the computer and software providers. This development will take place under the guidance of an information technology strategy produced by the organization's business managers.

A chief estimator needs to look at what the company is doing now and what it hopes to achieve in the future. He might ask himself these questions and put the answers in order of priority:

1. Why use computers?
   - To produce post-tender data.
   - To build up an accurate estimate of cost.
   - To allow tender adjustments to be made to an estimate, conveniently and accurately.
   - To reduce manual calculations.
   - To store standard models of bills and cost plans.
2. What are the basic needs?
   - Flexibility with different tender documentation.
   - Flexibility for projects with different time scales.
   - Hardware/software compatibility.
   - Uniformity of reports for basic resources, review meetings and post-tender allowances.
3. How do I select a system?
   - Attend a sales exhibition.
   - Ask for technical literature.
   - Find out what the estimators want.
   - Tell a supplier what you want.
   - Consult an independent expert.
   - Speak to other chief estimators.
4. What is *not* wanted?
   - Long, meaningless item code numbers.
   - Too many menus to change resources.
   - Long recalculation times.
5. What *is* wanted?
   - Company-specific reports.
   - Windows.
   - Context-sensitive help windows.
   - Rapid editing of bill items and prices.
   - Laser printer support.
   - Checking for unpriced items.
   - Clear reference manual.

One of the most difficult decisions is the scope of implementation. Should all estimators be introduced to computers at the same time? Can there be a combination of general-purpose and specialist packages? No one solution will suit all organizations. A cautious, flexible and evolutionary approach is the favoured solution. There will be some estimators who readily take to computers with the minimum of help, others might need training which can be conducted in-house or at a local college. A general understanding of desktop computing is a valuable skill which is beginning to appear in new entrants to the industry. Estimators have not been slow to recognize the benefits of computers as an aid to their work but have been cautious about the systems offered in the early days of personal computing.

The future

Computers are already an integral part of the estimating process. They help with the calculations and analysis of estimates, and produce clear reports for the construction team where the tender has been successful. Computer systems for estimating will gradually develop with the introduction of better hardware and user-friendly software. The first significant change will happen when the basic data (drawings, schedules or bills of quantities) are communicated to the contractor electronically or on disk. The two obstacles are training and compatibility of consultants' and contractors' systems.

Training is as important for the client's advisers as it is for estimators. In the early days of computing, students were expected to write simple programs which would solve problems which a spreadsheet would do today in a fraction of the time. With hindsight this was as useful as a van driver learning how to design a petrol engine. Now that general-purpose software is available in most offices, the most successful developments are where people's enthusiasm is channelled to speed up the operation of the company's procedures. Since computing skills are better taught through practical examples, training can take place at work without the need for time off at college. Clearly all estimators must have some computing skills and should feel comfortable using general-purpose software.

Both quantity surveyors and estimators have made a start but have approached the tender process from different standpoints. Quantity surveyors are naturally concerned with describing and quantifying the items of work for a project and the contractor needs to attach resources and prices to it. So quantity surveying software is designed to handle lists of work items in a structured form, whereas the estimator uses lists of resources which he can assign to the work. Coding methods have been tried but no common numbering system has emerged for general use and people resist codes as a way of entering and finding items. The answer might be for consultants to list activities using database software, and the database program would be distributed with the data. The contractor could then select items for sending to sub-contractors and price the rest of the work without the need for his own software. The main benefits would be:

1. The time saved by the contractors entering bills of quantities into their computers.
2. Amendments could be sent on disk or by electronic mail.
3. Priced bills could be returned with the tender in the form of a database file which would be transferred to a spreadsheet for checking and analysis of rates, and cashflow predictions.
4. The database file would be used for valuation purposes during construction.

There are several questions for the future which are constantly raised:

1. *Will computers replace estimators?* There have not been any reports of computers successfully replacing estimators but the information produced by estimators can be more comprehensive, and complex changes can be made to the estimate before the tender is submitted. There are many duties of an estimator which computers cannot replace: in particular, the many decisions made at each stage of a tender, site visits, discussions with sub-contractors, interpretation of ground conditions, access restrictions, best use of resources, and so on. Clearly the role of the estimator and his assistant is changing to adapt to the use of computers (see Fig. 20.13). The estimator will have time to produce more estimates or look in greater depth at the methods and resources for the contracts he wants to win.
2. *Are computer systems appropriate for operational estimating?*
    At the heart of most computer-aided estimating systems is the facility to price individual items of work, as listed in the bill of quantities. This method is well understood for building work but is difficult to use when a group of items must be priced as a single operation. On the other hand, a computer package will produce resource summaries which can be compared with a resourced programme. For example, the computer might show that an excavator has been allowed for one and a half weeks on a contract which clearly needs an excavator for two weeks. This information was difficult to draw out of the pricing notes using manual methods.
3. *How will expert systems aid the estimator?* In simple terms, an expert system can be described as having three parts: a user interface, a knowledge base (containing facts, rules and questions) and an inference engine which can draw on the knowledge base to make deductions about a particular problem. There has been a rapid growth in the number of off-the-shelf expert system shells designed for the IBM PC. A shell, comprising the user interface and inference engine, enables the user to input the knowledge part of a system.

273

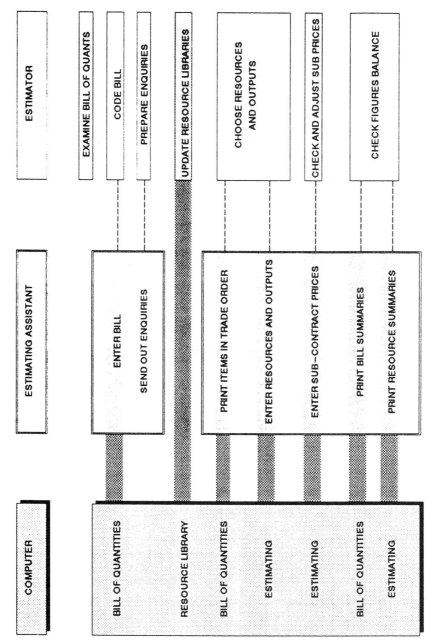

Fig 20.13 *The changing roles of the estimator and estimating assistant using computer systems*

274

In estimating, expert systems have been developed for early cost planning of industrial and commercial buildings. A client or his professional adviser can build up a cost model of a proposed building by answering a series of preselected questions, in a similar way to the many fault-diagnosis systems used in engineering and medicine. This approach could be used for some of the important decisions made by estimators and managers. The need for a tower crane, for example, must be proven before the costs are added to an estimate. The choice between scaffolding and hydraulic platforms is another question which can significantly affect the cost of temporary works. There are many applications which will be developed by contractors to suit their individual procedures, and will not be available on the open market, such as the setting of a margin to be added to an estimate.

It is unlikely that expert systems will replace general-purpose and estimating software. Estimators will continue to use spreadsheets and databases to schedule and price construction work.

4. *What do estimators want?* Advances in technology will please estimators when computers offer faster recalculation times for database systems. CD-ROM discs will be used to look up supplier and sub-contractor details and price lists, bills of quantities will be received electronically or on disk and tenders will be submitted in the same way. Journals will print independent test reports of estimating software, and eventually voice-recognition systems will allow estimators to talk to their computers, thus dispensing with the typewriter keyboard and other electronic input devices.

Clearly, the implementation of computers to aid the estimating function has not been easy. In fact there is evidence that some national contractors have abandoned their early attempts at introducing estimating systems and now encourage local experimentation with low-cost software.

There are other examples of contractors using estimating software to produce site budgets only on those tenders which are successful, that is, after the tender has been submitted. This is not so absurd as it seems. The effort required to get the bill of quantities in the computer is much more worth while if the bid is successful and every part of the tender can be allocated to sub-contract or direct work packages. All the adjustments made during the adjudication meeting can be incorporated into the costs and clear printouts can be produced for the construction team.

It is perhaps ironic that many of the industry's commentators greeted the last two decades of the 20th century as the time when the construction industry would see an end to the use of bills of quantities and the emergence of computer systems which could replace much of the work of estimators. We now know that both these predictions were wrong. Bills of quantities remain in everyday use and it is not possible yet to have a clear computing strategy for estimating.

Estimators will continue to develop their computing skills but they will not have one computer system which will meet all their needs. Experience now tells us that computers will not replace estimators; the latter will always be needed to predict the costs of construction with a reasonable degree of accuracy. What might change is their name and in some instances their status. With the drive towards greater economy, some estimating duties will be carried out by clerks, assistants and specialist buyers.

Further reading

Aqua Group (1990) *Tenders and Contracts for Building*, 2nd edition, Blackwell

Aqua Group (1992) *Pre-contract practice for the building team*, 8th edition, BSP

Ashworth, A. and Skitmore, R.M. (1982) Accuracy in estimating, Occasional Paper No. 27, The Chartered Institute of Building

Buchan, R.L., Fleming, F.W. and Kelly, J.R. (1991) *Estimating for Builders and Quantity Surveyors*, Butterworth-Heinemann

Chartered Institute of Building (1978) *Information required before estimating: a code of procedure supplementing the code of estimating practice*

Chartered Institute of Building, (1981) *The Practice of Estimating* (compiled and edited by P.Harlow)

Chartered Institute of Building (1983) *Code of Estimating Practice*, 5th edition

Chartered Institute of Building (1987) *Code of estimating practice*, Supplement No. 1, *Refurbishment and modernisation*

Chartered Institute of Building (1987) *Code of estimating practice*, Supplement No. 2, *Design and build*

Chartered Institute of Building (1987) *Code of estimating practice*, Supplement No. 3, *Management contracting*

Civil Engineering Standard Method of Measurement: 3rd edition (1991) Thomas Telford

Co-ordinating Committee for Project Information (1987) *Co-ordinated Project Information for Building Works, a guide with examples*, (this and other CPI documents may be obtained from BEC Publications, RIBA Publications or Surveyor's Bookshop)

Cook, A.E. (1991) *Construction Tendering: Theory and practice*, Batsford

Cross, D.M.G. (1990) *Builders' Estimating Data*, Heinemann-Newnes

Department of the Environment PSA (1987) *Significant Items Estimating* (produced by the Directorate of Building and Quantity Surveying Services), HMSO

Farrow, J.J. (1984) *Tendering – an applied science*, 2nd edition, the Chartered Institute of Building

Franks, J. (1990) *Building Procurement Systems*, the Chartered Institute of Building

McCaffer, R. and Baldwin, A. (1991) *Estimating and Tendering for Civil Engineering Works*, 2nd edition, BSP

Ministry of Public Building and Works (1964) *The placing and management of contracts for building and civil engineering work* (Report of the committee, Chairman: Sir Harold Banwell), HMSO

Ministry of Works (1944) *The placing and management of building contracts* (Report of the committee, Chairman: Sir Ernest Simon), HMSO

Skitmore, R.M. (1989) *Contract Bidding in Construction: Strategic management and modelling*, Longman Scientific & Technical

Smith, R.C. (1986) *Estimating and Tendering for Building Work*, Longman

Standard method of measurement of building works: 7th edition (1988) BEC and RICS

Whitehead, G. (1990) *Guide to Estimating for Heating and Ventilating Contractors*, HVCA

The Chartered Institute of Building, Technical Information Papers, 1982-1991
Paper No. 7 Potter, D. and Scoins, D. (1982) Computer-aided estimating
Paper No.9 Holes, L.G. and Thomas, R. (1982) A general-purpose data processing system for estimating
Paper No. 11 Harrison, R.S. (1982) Practicalities of computer-aided estimating
Paper No. 15 Skoyles, E.R. (1982) Waste and the estimator
Paper No. 37 Harrison, R.S. (1984) Pricing drainage and external works
Paper No. 39 Braid, S.R. (1984) Importance of estimating feedback
Paper No. 59 Uprichard, D.C. (1986) Computerised standard networks in tender planning
Paper No. 64 Ashworth, A. (1986) Cost models – their history, development and appraisal
Paper No. 65 Ashworth, A. and Elliott, D.A. (1986) Price books and schedules of rates
Paper No. 75 Harrison, R.S. (1987) Managing the estimating function
Paper No. 77 Ashworth, A. (1987) General and other attendance provided for sub-contractors
Paper No. 81 Ashworth, A. (1987) The computer and the estimator
Paper No. 97 Brook, M.D. (1988) The use of spreadsheets in estimating
Paper No. 114 Emsley, M.W. and Harris, F.C. (1990) Methods and rates for structural steel erection
Paper No. 120 Cook, A.E. (1990) The cost of preparing tenders for fixed price contracts
Paper No. 127 Harris, F. and McCaffer, R. (1991) The management of contractor's plant
Paper No. 128 Price, A.D.F. (1991) Measurement of construction productivity: Concrete gangs
Paper No. 131 Brook, M.D. (1991) Safety considerations on tendering – management's responsibility

The Chartered Institute of Building, Construction Papers, 1992
Paper No. 2 Massey, W.B. (1992) Sub-contractors during the tender period – an estimator's view
Paper No. 11 Hardy, T.J. (1992) Germany – a challenge for the estimator

The National Joint Consultative Committee for Building (RIBA and BEC)
Code of Procedure for Single Stage Selective Tendering (1989)
Code of Procedure for Two Stage Selective Tendering (1983)
Code of Procedure for Selective Tendering for Design and Build (1985)
Code of Procedure for the Selection of a Management Contractor and Works Contractors (1990)
Guidance Note 1.   Joint venture tendering for contracts in the United Kingdom (1985)
Guidance Note 2.   Performance bonds (1986)
Guidance Note 3.   Fire officers' recommendations (1987)
Guidance Note 4.   Pre-tender meetings (1986)
Guidance Note 5.   Reproduction of drawings for tender purposes (1990)
Guidance Note 6.   Collateral warranties (1990)

Index

279